Springer Proceedings in Earth and Environmental Sciences

The series Springer Proceedings in Earth and Environmental Sciences publishes proceedings from scholarly meetings and workshops on all topics related to Environmental and Earth Sciences and related sciences. This series constitutes a comprehensive up-to-date source of reference on a field or subfield of relevance in Earth and Environmental Sciences. In addition to an overall evaluation of the interest, scientific quality, and timeliness of each proposal at the hands of the publisher, individual contributions are all refereed to the high quality standards of leading journals in the field. Thus, this series provides the research community with well-edited, authoritative reports on developments in the most exciting areas of environmental sciences, earth sciences and related fields.

André Podleisek · María A. Franco · Daniel Kliem ·
Emanuele Carpanzano
Editors

Advances in Circular Economy: Bridging Research and Practice

Proceedings of the FTAL Conference 2024 on Circular Economy

Springer

Editors
André Podleisek
Eastern Switzerland University of Applied Sciences (OST)
Rapperswil, Switzerland

Daniel Kliem
Eastern Switzerland University of Applied Sciences (OST)
St. Gallen, Switzerland

María A. Franco
School of Engineering and Computer Science
Bern University of Applied Sciences (BFH)
Biel, Switzerland

Emanuele Carpanzano
University of Applied Sciences and Arts of Southern Switzerland (SUPSI)
Manno, Ticino, Switzerland

ISSN 2524-342X ISSN 2524-3438 (electronic)
Springer Proceedings in Earth and Environmental Sciences
ISBN 978-3-032-21546-8 ISBN 978-3-032-21547-5 (eBook)
https://doi.org/10.1007/978-3-032-21547-5

This work was supported by the University of Applied Sciences and Arts of Southern Switzerland (SUPSI) and the Association of Swiss Schools of Engineering, Architecture and Life Sciences (FTAL).

This Springer imprint is published by the registered company Springer Nature Switzerland AG
The registered company address is: Gewerbestrasse 11, 6330 Cham, Switzerland

Editorial

Conference On the State of Research On Circular Economy

In recent years, the concept of circular economy has been widespread, leading a growing number of governmental, societal, and business stakeholders to take major initiatives to find promising cyclical models alternative to the currently widely used linear ones.

The adoption of circular economy is of paramount importance for reducing environmental impact, creating fair economic opportunities, and building valuable models capable of facing challenging global dynamics. The circular economy offers innovative solutions to face the current socioeconomic and environmental challenges and to create a more sustainable future.

Within this context, the Association of Swiss Schools of Engineering, Architecture and Life Sciences (FTAL) invited all researchers and scholars to the FTAL 2024 conference on Circular Economy on November 14 and 15, 2024 in Lugano (Switzerland). The conference on Circular Economy aimed to present the state of the art of research, and addressed topics such as resource efficiency, value retention, servitization, localization, circular built environment, circular bioeconomy and other new business, production, and market models.

The goals of the conference were as follows: first, to promote exchanges between the various disciplines encompassed by the Circular Economy, offering new insights and research challenges; second, to gather input from international experts to advance ongoing research; third, to connect practitioners and researchers in order to foster the development and application of grounded theories; and finally, to present the state of the art in circular frameworks and solutions.

The conference welcomed submissions from professors, researchers, Ph.D. and master students from different disciplines and practitioners. Three types of contributions were sought for: Full papers to share and discuss main results of the paper as well as work in progress and planned projects to get feedback on the approach and initial findings. In addition, the conference accepted proposals for interactive workshops of 30 or 60 min, where a defined target group of the conference was brought together to work within a specific methodology on a set of research questions and advance the project.

Steering Committee

Emanuele Carpanzano, SUPSI
Oya Atalay Franck, ZHAW
Olivier Naef, HES-SO
Falko Schlottig, FHNW
Viktor Sigrist, HSLU
Dirk Wilhelm, ZHAW

Organizing Committee

Corinna Koch Dandolo, SUPSI
Alice Noris, SUPSI
Esra Hergert, FTAL Fachkonferenz Technik, Architektur und Life Sciences

Scientific Committee

Chair: André Podleisek, OST
Vice Chair: Maria Franco Mosquera, BFH
Vice Chair: Daniel Kliem, OST
Norman Urs Baier, BFH
Corinna Baumgartner, ZHAW
Adrian Burri, ZHAW
Donatella Corti, SUPSI
Lorenzo Di Lucia, SUPSI
Alessandro Fontana, SUPSI
Elimar Frank, OST
Christine Grimm, HSLU
Aurelia Kogler, FHGR
Rolf Krebs, ZHAW
Markus Lenz, FHNW
Korbinian Schneider, FHNW
Daniel Schwendemann, OST
Hans Ulrich Siegenthaler, HES-SO
Marco Viviani, HES-SO
Stefano Zerbi, SUPSI

Focus Areas for Papers, Work-in-Progress and Interactive Sessions

Contributions address at least one of five topic areas:

(1) Innovative concepts for the practical implementation of the circular Rs (Refuse, Reduce, Reuse, Repair, Refurbish, Remanufacture, Repurpose, Recycle, Recover, Remine), especially those involving models for multiple or combined cycles.
(2) Business models and strategies to rethink and redesign towards circular value networks or localization, including costs, value retention, revalorization, and incentives.
(3) Societal changes, community approaches and motivations, including policy and regulatory frameworks.
(4) Technical challenges like reverse logistics, material durability, and contamination.
(5) Monitoring and assessment of circularity, including transparency, traceability, and its impact on the environment, society, and businesses.

The two-day conference, organized into four parallel sessions, consisted of five full-paper tracks, four work-in-progress tracks, and three workshop tracks. For each track, an overarching theme was defined, and the matching contributions were grouped accordingly.

Full Paper (FP) Tracks

FP Track 1—Circular Plastics Management

This track explores comprehensive approaches to plastic waste management from the perspectives of policy, public awareness, and stakeholder engagement. The focus is on developing circular economy solutions that address plastic pollution challenges, particularly in developing countries. Research examines policy instruments, implementation strategies, and recommendations for creating zero-waste models. The track emphasizes the importance of capacity building, infrastructure development, and stakeholder collaboration in achieving effective plastic waste management. Studies also investigate innovative business models that demonstrate how plastic circularity can serve as a competitive advantage while reducing environmental impact.

FP Track 2—Material Innovation and Recycling Technologies

This track centers on cutting-edge material innovations and advanced recycling technologies that enhance sustainability across various industries. Research focuses on developing composite materials from recycled sources up to high-performance applications. The track explores innovative joining techniques and design-for-disassembly approaches that maximize material efficiency and enable circular material flows. Studies also examine the integration of biological materials into mass-produced products, demonstrating how sustainable alternatives can replace conventional materials while maintaining functional performance. The track emphasizes the technical feasibility and environmental benefits of advanced recycling processes.

FP Track 3—Circular Construction and Design

This track addresses the transformation of the construction industry toward circular practices through innovative structural systems and sustainable design approaches. Research examines the reuse of concrete elements and other building components in new construction projects, focusing on design-for-disassembly principles that enable future material recovery and reuse. The track explores the development of assessment tools and frameworks for measuring circularity in buildings and neighborhoods. Studies investigate fire-resistant composite systems that reduce embodied carbon while maintaining structural integrity. The track emphasizes the integration of circular principles into building design processes and the development of practical tools for sustainable construction.

FP Track 4—Business Performance and Circular Practices

This track investigates the relationship between circular economy implementation and business performance across various industry sectors. Research examines how companies that embed circular economy principles into their core operations experience different performance outcomes compared to those implementing circular practices as separate initiatives. The track also explores innovative financing mechanisms at the regional level. Studies analyze measurement methodologies for assessing circularity within specific industry contexts. The track emphasizes the economic viability and competitive advantages of circular business models while addressing implementation challenges and opportunities.

FP Track 5—Circular Economy in the Fashion Industry

This track focuses on the transformation of the fashion industry toward circular practices through innovative approaches to textile waste management and sustainable production. Research examines the development of biologically circular textiles, and the challenges associated with industrial composting of textile materials. The track explores advanced recycling technologies for post-industrial textile waste, particularly addressing the complex issue of mixed fiber recycling. Studies investigate regional case studies that assess the maturity of circular economy adoption within fashion manufacturing companies. The track emphasizes the need for regulatory changes, stakeholder collaboration, and technological advancements to enable effective circular practices in the fashion sector.

Work in Progress (WiP) Tracks

WiP Track 1—Circular Construction and Design

This track presents ongoing research into holistic frameworks for circular construction that address current trends, system needs, and sustainable objectives. Projects examine the reuse and remanufacturing of construction wood, exploring cascading utilization strategies that maximize material value while minimizing forest impact. Research investigates the development of low-carbon binders using local waste materials that reduce cement content while maintaining structural performance. The track focuses on addressing gaps in current construction practices, including the limited integration of circularity metrics into lifecycle assessments and the need for automated decision-making criteria in construction projects.

WiP Track 2—Circular Economy in Agro-Industrial and Farm Settings

This track explores innovative approaches to valorizing agricultural and industrial waste streams through circular economy principles. Research examines the extraction and application of high-value materials from agricultural byproducts. Projects investigate pyrolysis technologies for biomass conversion. The track also addresses lifecycle assessment methodologies for coffee packaging options. Studies emphasize the potential for agricultural waste streams to contribute to sustainable production systems while reducing environmental impacts.

WiP Track 3—Business and Innovation Models in Circular Economy

This track examines emerging business models and innovation strategies that enable circular economy transitions across different sectors and regions. Research investigates venture capital funding patterns for circular economy startups. Projects also explore the development of living labs in rural areas. They examine innovative service models, analyze how digital platforms are reshaping second-hand markets and influencing consumer behavior toward sustainable consumption patterns.

WiP Track 4—Energy, Material and Waste Sustainability Solutions

This track addresses comprehensive approaches to energy efficiency, material recovery, and waste management through innovative technological solutions. Research examines lifecycle assessment methodologies specifically tailored to the plastics sector, addressing the unique challenges and complexities of environmental impact assessment in this industry. Projects investigate the development of bio-based hydrogels from recycled polymer sources for wastewater treatment applications. Studies also evaluate the contribution of small-scale, decentralized biogas plants. The track explores end-of-life strategies for emerging technologies, focusing on the recovery of critical materials.

Workshop (WS) Tracks

WS Track 1.1 and 1.2—Building Reuse Ecosystem

These interconnected workshops focus on developing and strengthening the building reuse ecosystem through stakeholder engagement and platform development. The sessions explore the creation of digital marketplaces for building components that facilitate the exchange of reused construction materials at the regional level. Workshops address the unique challenges of implementing circular processes in the construction industry, including acceptance levels for reused components and the benefits and criticalities of exchange platforms. The workshops emphasize practical solutions for overcoming communication gaps, logistical challenges, and economic barriers that currently limit the adoption of building component reuse.

WS Track 2—Business Model Transformation

This workshop provides practical frameworks and tools for companies transitioning to circular economy business models. The session covers core elements essential for circular transformation, including circular design principles, business model innovation, supply chain optimization, and logistics strategies. The workshop aims to equip participants with actionable tools and techniques for implementing circular strategies within their organizations while fostering innovative, solution-oriented thinking.

Findings from the Keynotes

David Peck: Critical Materials and Circularity in the Swiss Context: Navigating Resource Constraints Through Circular Strategies

In his keynote, Dr. David Peck from Delft University of Technology offered a compelling examination of critical raw materials (CRMs)[1] from a global and Swiss perspective, highlighting the urgent need for circular economy approaches to address resource constraints. Beginning with a historical parallel between the material scarcity faced between 1939 and 1943 and contemporary challenges, Dr. Peck framed a provocative question: "If you had to halve materials and energy use in Switzerland, how could you do it and how quickly?". He then relied on historical examples, such as the British response to material shortages during WWII, to highlight how government intervention and centralized control enabled efficient, resource-conserving production. He noted many of the principles applied back then, such as prioritizing essential goods, designing for material efficiency and durability, and promoting repair and reuse, align with today's circular economy ideals, even if they were not intentionally "circular" at the time.

Dr. Peck then proposed an approach to the management of critical raw materials that is holistic and interdisciplinary, which integrates engineering and design, business, policy and regulation, as well as societal values. In connection to product design, he noted that while supply risks and price volatility are recognized as threats to the availability of CRMs, these concerns rarely translate into design activity within companies. Although product design is recognized as critical in the circular economy transition today, it is an overlooked design driver for addressing critical materials challenges. He, therefore, made a call not only to consider the criticality and recyclability of materials in product design (besides functionality and cost), but also to develop substitute and improved materials.

[1] In the context of EU strategy, a critical raw material (CRM) is defined as a raw material that is both of high economic importance to the EU and at high risk of supply disruption due to factors such as concentrated sources, lack of substitutes, and geopolitical dependencies.

When it comes to businesses, Dr. Peck called for a shift in business models towards circularity, where value is created through longevity, reuse, and service-based approaches, rather than linear product consumption. It was also argued that while today's transitions are largely private sector-led with government encouragement, there is a need for stronger, more centralized support and policy measures to accelerate change and ensure resilience. Examples of such interventions are: government-supported materials data repositories, country/regional plans to address material shortages, taxes, and incentives for widespread circular public sector procurement. Furthermore, Dr. David Peck shared his view of the supply and demand of critical raw materials (CRMs) as being an inherently global challenge, shaped by complex, interconnected supply chains and escalating geopolitical tensions. He raised a call for coordinated action both at the national and international levels to build resilience and diversify supply.

A central theme in Peck's keynote was the need to move from a focus on individual product consumption (with little to no knowledge about materials' value and true costs) to a broader societal goal of well-being and sustainability. He pointed out that during times of crisis (e.g., WWII), societies prioritized essential needs and collective well-being, resulting in resource-efficient and durable products, a lesson he sees as highly relevant today.

In his keynote, Dr. Peck also addressed the current state of critical raw materials management in Switzerland. Switzerland currently lacks its own CRM methodology and list, relying instead on EU data for strategic planning. Despite ongoing discussions in the Swiss Federal Council regarding the importance of CRMs and dependent technologies, Swiss policy focuses primarily on Rare Earth Elements (REE) while failing to establish strong links between CRM supply and circularity principles. This represents, according to Dr. Peck, a significant gap in the country's approach to resource security, particularly as Switzerland faces similar demand patterns to the EU but occupies a more vulnerable position in terms of supply chain resilience.

The question of whether secondary material supply could meet current and future demand, was also discussed during Dr. Peck's keynote. His position on the continued extraction of both primary (i.e., mining) and secondary (i.e., recycling and urban mining) materials was pragmatic and grounded in the realities of current technological and societal needs. He recognized that the green and digital transitions, including renewable energy technologies, batteries, and electronics, as well as sectors like aerospace and defense, depend on a range of critical raw materials that cannot be substituted or eliminated in the short term. Peck's overarching view was that while the extraction of primary materials cannot be eliminated in the near future, it must, to the greatest extent possible, be driven by less polluting methods and stricter environmental standards. In his view, primary extraction does need to be complemented by secondary sourcing as part of a transition to a circular economy. Secondary material extraction through recycling and urban mining, however, also presents its hurdles. Challenges come from the fact that most materials are still locked in products and infrastructure in use, and only become available for recycling at the end of their life cycles. Circular approaches that promote extension of product lifecycles, such as remanufacturing, contribute to the delays in material availability for recycling as well. Finally, technological, economic, and logistical barriers limit the efficiency and profitability of recovering critical elements, especially those present in small quantities or complex forms, making it impossible for secondary sources alone to match the scale and mix of materials society requires.

The presentation concluded with practical recommendations for Switzerland: reduce demand through research and innovation in materials; diversify technology mixes to limit reliance on specific materials; establish robust recycling infrastructures; develop strategies for extended product life; support social innovation around sharing and reuse; and enhance waste prevention. Finally, Dr. Peck urged Switzerland to work collaboratively with the EU on supply chain innovations, creating a detailed overview of critical materials needed for the energy transition, and establishing secure supply chains moving forward.

Melanie Haupt: Advancing Circular Economy in Switzerland: Bridging Theory, Practice, and Policy

Dr. Melanie Haupt's keynote was a profound reflection on the essential intersection between research, industry, and regulation in advancing the circular economy (CE). In her address, Dr. Haupt underlined the importance of putting circularity into practice, transforming theoretical frameworks into actionable solutions to mitigate environmental degradation. She presented a compelling narrative emphasizing the need for cross-sectoral collaboration and scientific innovation to achieve a truly sustainable circular economy.

A central theme of Dr. Haupt's keynote was the definition and scope of circular economy, particularly its regenerative nature. Circular economy, as she highlighted, is not merely a process of recycling or reusing materials, but a holistic economic system that requires a fundamental shift from the traditional 'end-of-life' model. The approach seeks to preserve resources, foster value retention, and promote sustainable development across all stages of product life cycles. Dr. Haupt effectively illustrated the complexity of this concept by referencing the work of Kirchherr et al. (2023), which identified over 220 distinct definitions of circular economy. This plurality of definitions underscores the challenge of establishing a universally accepted framework, yet it also signals the dynamism and relevance of the field in contemporary environmental discourse.

Drawing from the Circularity Gap Report of Switzerland (2023), she also illustrated the stark imbalance between global material flows and the actual rate of recycling and reuse. This gap represents not only an environmental challenge but also an economic one, as it reflects the untapped potential for circular business models to enhance resource security and reduce dependency on fragile supply chains. For Switzerland, a nation known for its robust environmental policies, the challenge is particularly significant. Dr. Haupt emphasized that scientific advancements in circular practices and clear policy directives are crucial for transitioning towards a circular economy at both national and global levels.

Dr. Haupt also explored the nuanced relationship between circularity and sustainability, noting that while circular economy principles align with sustainability goals, they do not automatically ensure environmental benefits. She cited specific examples where practices like remanufacturing, despite their circular nature, could lead to unintended negative environmental consequences. Hence, circularity does not guarantee sustainability unless the environmental impact of each circular practice is carefully assessed and optimized. This distinction calls for a deeper, more holistic approach to circular economy strategies that integrates ecological, economic, and social dimensions, ensuring that the shift towards circularity does not inadvertently exacerbate other forms of environmental harm.

A vital part of Dr. Haupt's presentation was the focus on business models in the context of circular economy. She argued that businesses are central to the circular economy transformation, positioning them not only as agents of change but also as beneficiaries of circular strategies. Companies that adopt circular business models can not only reduce costs and supply chain risks but can also enhance their market resilience and long-term profitability. Drawing on examples such as the transition from linear models to circular product-as-a-service models, Dr. Haupt highlighted how innovative business designs can close material loops, facilitate resource sharing, and create new value streams. She made a compelling case for the reimagining of product lifecycles, from design through reuse, remanufacturing, and recycling, to reflect a more sustainable and economically viable approach.

Moreover, Dr. Haupt's discussion on the role of regulators was particularly pertinent, as she identified the need for robust legal frameworks to facilitate the transition to circular economies. Her analysis of the recent revisions to the Swiss Environmental Protection Act, along with the upcoming EU regulations, including the Ecodesign for Sustainable Products Regulation and the EU Strategy for Sustainable Textiles, revealed the increasing recognition of circularity in public policy. These regulatory advancements emphasize not only product durability and recyclability but also the environmental responsibility of producers through extended producer responsibility (EPR) schemes. Dr. Haupt emphasized that while regulations are vital, they must be measurable and meaningful, guiding industries towards tangible outcomes in resource conservation and environmental protection. All in all, Dr. Haupt's address was a powerful reminder that the transition to circularity is not just about technological solutions, but about rethinking how we design, consume, and think about resources in the modern world.

Maria Castillo: Financing the Circular Transition: Strategic Imperatives for Swiss Financial Institutions

Maria Castillo's address provided attendees with a much-needed financial perspective on circularity—one that complements the technical and material innovation narratives prevalent throughout the conference. The keynote began with a sobering assessment of the global circularity gap: despite heightened awareness of circular principles, the share of global materials flowing through circular systems has dropped from 9% in 2018 to just 7% in 2023. Ms. Castillo contextualized this regression by pointing to a lack of scalable financing frameworks, limited standardization, and insufficient awareness among financial institutions regarding circular business models. She emphasized that if the circular economy is to become more than an aspirational concept, the financial sector must view circularity not only as an ESG consideration but as a long-term economic imperative.

At the heart of Ms. Castillo's presentation was a meticulous examination of the regulatory landscape shaping circular finance in Europe. Drawing from foundational instruments such as the EU Circular Economy Action Plan, the EU Taxonomy for sustainable activities, the Sustainable Finance Disclosure Regulation (SFDR), and the Corporate Sustainability Reporting Directive (CSRD), she outlined a growing web of obligations and incentives aimed at aligning financial markets with circular objectives. Notably, she illustrated how these frameworks are progressively redefining risk, transparency, and fiduciary duty, urging financial actors to integrate circularity as a core strategic consideration.

Beyond the legislative narratives, Castillo addressed the expectations placed on financial institutions in operationalizing the circular transition. From setting sector-specific carbon and resource efficiency targets to developing bespoke financing instruments, such as green loans and sustainability-linked bonds, financial institutions are increasingly expected to fund companies that embed circularity into their value chains. She presented the UNEP FI's guidance on embedding circularity in climate transition plans, underscoring that financial institutions must not only adapt to external regulations but must also proactively shape capital flows that reflect a circular ethos.

A particularly insightful portion of the keynote was devoted to the presentation of findings from ECOFACT's peer benchmarking assessment of 17 leading international financial institutions. The analysis revealed that while all institutions made reference to the circular economy in their policies, few integrated it as part of their risk management processes. Most framed circularity as an opportunity rather than a mitigation strategy, often in the context of supporting client transitions or launching thematic investment products. This duality, between opportunity and obligation, captures the state of strategic uncertainty in which financial institutions are balancing legacy practices with the emerging expectations of a circular economy. Ms. Castillo deftly argued that full alignment with circular economy goals requires embedding circularity not just into investment strategies but also into the fundamental frameworks of risk assessment, client engagement, and value chain evaluation.

Her presentation also delved into sectoral priorities, identifying industries such as construction, real estate, manufacturing, packaging, and ICT as key arenas where financial support can accelerate circular outcomes. Castillo emphasized the importance of value chain thinking, where upstream, midstream, and downstream actors must be accounted for in financing decisions. This systemic view of financing aligns with the notion that everyone's output becomes someone else's input, a principle foundational to circularity.

In conclusion, Ms. Castillo summarized six key takeaways, among them the pressing need to integrate circular economy principles into double materiality assessments, and the necessity of leveraging regulatory momentum to enhance transparency and data-driven decision-making. Perhaps most compelling was her argument that circular economy financing should not remain in the periphery of sustainability discourse, but be recognized as a driver for achieving long-term climate, economic, and societal resilience.

Maria Castillo's keynote was a timely and expertly crafted intervention, reminding us that the realization of a circular economy cannot rest solely on designers, engineers, or policymakers. Capital flows shape the architecture of our economy, and unless finance is both mobilized and reimagined through the lens of circularity, progress will remain incremental. Her contribution to FTAL 2024 therefore challenged us to view financial institutions not as passive enablers but as active architects of systemic change in a circular society.

Recognition and Awards

Outstanding contributions were selected by the members of the scientific committee in a structured review process.

The following categories were awarded at the end of the conference:

- Best Scientific Paper: Marco Ricchetti, Aurora Magni, Alessandro Fontana, Deborah Leone and Ludovica Rossi for "Advancing Circular Economy in Fashion: Feasibility and Impact of Enhanced Recycling Technologies for post-industrial Textile Waste".
- Best Interdisciplinary Project: Line Fournier, Larissa Hänni and Maria Franco Mosquera for "Applying a transdisciplinary approach to the development of a biologically circular beanie".
- Best Poster Contribution: Roger Vergés for "A Holistic Framework Proposal for Circular Construction from Current Trends, System Needs, and Sustainable Objectives".
- Best Interactive Session: Elimar Frank, Tom Koch and Frédéric Michaud for the workshop "Core Elements for the Transition of Companies to Circular Economy".

Special recognition was given to the best Master's or Ph.D. thesis addressing a technological innovation with societal benefits and positive impact.

- "Technology for Society Award": Leidy Guante Henriquez for "From Reviews of Sustainable Building Rating Systems and Circular Indicators for the Built Environment to a Practical Tool Supporting the Design of Sustainable Circular Buildings".

Rapperswil, Switzerland
André Podleisek
andre.podleisek@ost.ch

Biel, Switzerland
María A. Franco
maria.francomosquera@bfh.ch

Contents

Holistic Transformation: Integrating Sustainability in Banking through Crowdfunding

Alessandra Versi

Abstract This study investigates sustainability-oriented crowdfunding as a tool for financial institutions to promote sustainable entrepreneurship in Ticino. Using stakeholder insights, expert interviews, focus groups, pilot survey and Business Model Canvas, the research identifies opportunities and challenges in aligning crowdfunding initiatives with sustainability goals. The proposed sustainability-focused crowdfunding platform in Ticino aims to enhance local engagement in sustainability initiatives through a thorough stakeholder analysis. This analysis reveals a diverse range of actors, including institutional organizations, as well as private investors and businesses, each offering unique strengths and varying commitments to sustainability. A bottom-up approach is suggested to counter the perception of sustainability as merely an image-building exercise, focusing on local identity and culture to foster genuine community involvement. However, challenges such as resistance to collaboration due to a silo mentality must be addressed by promoting inclusive and effective communication strategies. Integrating the platform within Cantonal bank's operations presents both opportunities for increased credibility and challenges in aligning with the bank's evolving CSR strategy. Ensuring user-friendly access and optimizing project selection based on sustainability criteria are vital for long-term success.

Keywords Sustainability-oriented crowdfunding · Sustainable entrepreneurship · Sustainable finance

1 Introduction

Sustainability in entrepreneurship is increasingly viewed as a key driver for achieving sustainable development, addressing present needs without compromising future generations' ability to meet theirs [1, 2]. However, sustainable entrepreneurs face

---

A. Versi (✉)
School of Agricultural, Forest and Food Sciences, Bern University of Applied Science, Zollikofen, Switzerland
e-mail: alessandra.versi@hotmail.com

A. Podleisek et al. (eds.), *Advances in Circular Economy: Bridging Research and Practice*, Springer Proceedings in Earth and Environmental Sciences,
https://doi.org/10.1007/978-3-032-21547-5_1

challenges such as limited business experience, scarce financing, and significant social and environmental constraints [3]. Furthermore, they are often hindered by the "lock-in effect," which limits transitions from traditional "business as usual" models to innovative, sustainable alternatives [4].

Crowdfunding has emerged as a promising solution, leveraging digital platforms to fund projects aligned with environmental, social, and governance (ESG) principles [5, 6]. This approach not only supports sustainable development but also fosters community engagement, aligning initiatives with local values and needs [7]. For financial institutions, such as cantonal banks, integrating sustainable practices is critical to addressing national and global sustainability priorities [8].

This study explores the feasibility of establishing a sustainability-focused crowdfunding platform in Ticino. The research is structured around three key questions:

1. What are the challenges and opportunities in integrating sustainability-focused crowdfunding?
2. How can crowdfunding support a financial institution's sustainability goals?
3. What criteria are essential for selecting projects to be funded?

Supported by the Cantonal bank of Ticino in its role as an external expert, the research combines stakeholder analysis, regional context exploration, and project evaluation criteria, aiming to create a framework that fosters synergy between sustainability, innovation, and community engagement. Conducted as part of the Master of Science program in Circular Innovation and Sustainability at the University of Applied Science of Bern (BFH), this study provides actionable insights for the banking sector.

2 Methodology

Data collection involved five semi-structured interviews and a focus group session, all conducted in Italian and later transcribed into English. The interviews were carried out with a diverse group of experts, including a Swiss Triple Impact coach, a crowdfunding platform manager, an angel investor, a cantonal micro-entrepreneur, and a social entrepreneur. The data gathered was coded, culminating in the creation of a comprehensive risk and opportunity map.

The focus group, lasting 1.5 h, included a regional development officer, a project promoter, and a digital infrastructure expert. Participants positioned actors based on proximity to the platform and assigned roles for platform-related activities. In a second phase, connections were made to identify potential contributions from non-assigned actors, concluding with an open discussion on platform opportunities and challenges.

The study also employed Business Model Canvas (BMC) to evaluate the cantonal bank's (BancaStato) business model. It analyzed the bank's mission, vision, target customers, value proposition, distribution channels, customer relationship strategies, and key resources and partnerships. A case study of Raiffeisen Bank's crowdfunding platform [9] was conducted using BMC, identifying best practices for integration into the cantonal bank's model.

Further methods included a literature review on project selection in crowdfunding and a competitive benchmark of user experience to assess selection practices. A pilot survey, conducted with 22 participants over three weeks, tested the platform's landing page for clarity, search functionalities, and user engagement. It also evaluated whether external promotions influenced users' sustainability knowledge and how much time new users were willing to invest on the platform.

3 Results

3.1 Stakeholders' Analysis

The stakeholder map (→ Fig. 1), developed from focus group findings, identifies key actors in sustainability initiatives in Ticino. These stakeholders can be categorized based on their engagement with sustainability and their contribution to the platform.

Institutional stakeholders such as Cantonal entrepreneurship support initiatives (e.g. Fondounimpresa and ImpactHub), which support entrepreneurs and impact-driven projects, are central. They can help promote and evaluate sustainability initiatives. Additionally, economic organizations, like Bank institutions (e.g. BancaStato) and the Chamber of Commerce are important for project funding and promotion.

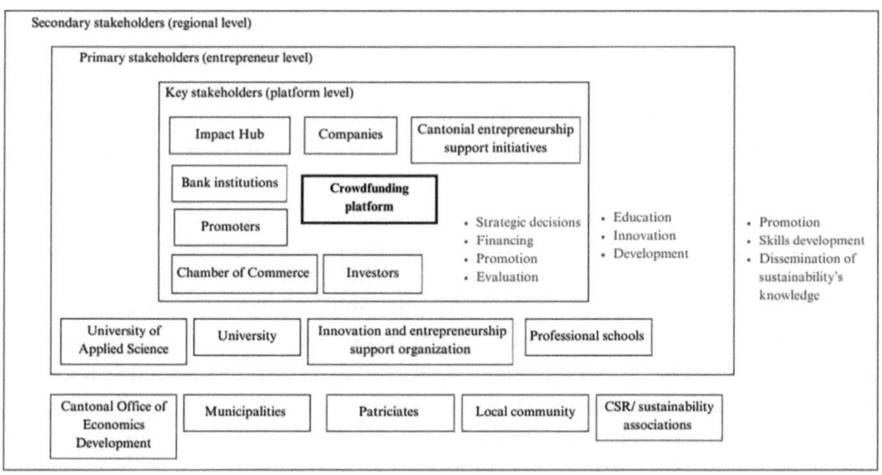

Fig. 1 A stakeholder analysis diagram depicting the categorization of key, primary, and secondary stakeholders based on their roles, influence, and activities. The diagram visually organizes stakeholders into three distinct groups, highlighting their respective contributions to the project's success.

Private stakeholders, including Ticino-based businesses, project promoters, and investors, are crucial for implementing sustainable initiatives and securing necessary resources.

Primary stakeholders seem to focus more on education, supporting innovation, and regional technological development to aid SMEs and innovative startups in Ticino. At the level of secondary stakeholders, public entities such as the Canton Office of Economic Development municipalities, and patriciates are classified, along with local communities and industry associations. These entities may engage in project promotion and ideation, skills development, training, as well as communication and dissemination activities for sustainable initiatives.

3.2 Opportunities and Challenges of the Integration of Sustainability-Focused Crowdfunding

The project identified opportunities and challenges within the ecosystem. In Ticino, sustainability efforts by the government, especially through the Economic Division, are recognized: "[…] the Canton is investing significantly in this area" [Interview 4]. However, sustainability is still new, requiring further sensitization and community engagement: "[…] they do not see the economic return in the short term, as sustainability results tend to be in the long term" [Interview 1]. There is resistance to viewing sustainability as profitable, and a top-down approach dominates. Shifting to a bottom-up strategy could enhance initiatives and foster community participation, vital for the platform's success.

The platform could help organizations transition gradually, leveraging crowdfunding to yield tangible short-term results while overcoming challenges like resistance to collaboration: "[…] some projects might decide to seek funding outside the Canton of Ticino, such as going to Zurich" [Interview 2]. A sustainability-oriented approach demands inclusivity: "[…] everyone's involvement is essential for positive outcomes" [Interview 1].

Strong partnerships are key for the platform's credibility and success. It must actively engage stakeholders to create an aggregative ecosystem: "[…] platform shouldn't solely focus on fundraising but should foster a sense of aggregation" [Interview 4]. Clear roles for partners and collaboration with local entities, including social entrepreneurs, can strengthen the platform. Businesses are increasingly interested in sustainability projects: "[…] companies are forced to be responsible and sustainable, especially by public institutions" [Interview 5].

To attract partners, a benefit plan is essential, offering incentives like visibility, advertising, or market testing opportunities: "[…] a good narrative and concrete incentives need to be found" [Interview 5].

Awareness of crowdfunding is limited, and promoters often underestimate the effort required: "[…] project promoters often have vague ideas about the business economics part of the project" [Interview 2]. The platform can leverage its local expertise to differentiate from larger, less region-specific competitors. Trust is crucial, requiring transparency and coherence between mission and activities: "[…] transparency is a fundamental factor, because investors want to know where their money is allocated" [Interview 3].

Finally, a rigorous project selection process ensures credibility while balancing the fragility of early-stage projects: "[…] accurate project selection is important to maintain platform credibility" [Interview 1]. Success stories are vital for visibility and funding, highlighting the need for structured promotion plans: "[…] the most useful thing is visibility. Therefore, the platform having a large community is crucial" [Interview 5].

3.3 Integration of a Sustainability-Oriented Crowdfunding Platform

A Cantonal bank is committed to driving the Canton's economic growth while adhering to its public mandate. The bank plays an active role in wealth redistribution and maintains a strong local presence. Its vision and mission emphasize becoming a key reference point for local development, enhancing the regional economy's added value, safeguarding customer savings, and offering attractive savings products [10].

BancaStato combines physical and digital sales networks, prioritizing customer relationships to inspire trust in financial management. It maintains significant partnerships with entities like the Republic of Canton Ticino and other Regional Innovation System players, underlining the importance of local connections [11].

The bank's customer base primarily comprises private individuals and local SMEs, reflecting its deep integration within the Canton. This demonstrates its role not only in financial services but also in actively supporting local development. By delivering tailored services to meet community needs, BancaStato upholds principles of social and environmental responsibility outlined in its public mandate [12].

Despite the range of customer relationship opportunities, this relationship remains mostly passive. While feedback mechanisms exist, there is limited scope for clients to take the initiative. Consequently, the lack of active involvement prevents the formation of a client community.

Although BancaStato has established itself as Canton's trusted financial institution for over a century, new opportunities arise for fostering community engagement through a sustainability-focused platform. The bank's relatively young CSR strategy offers flexibility to integrate such a platform into its business model effectively.

There are operational improvements that a cantonal bank should undertake, which would allow it to address the challenges for the implementation of a sustainable crowdfunding platform.

Currently sustainability and the associated approaches to it are still poorly structured, applying standardized approaches is complex due to the holistic and peculiar nature of the latter. Furthermore, as previously discussed, awareness of sustainable practices in the market is still limited. Therefore, it is still difficult to accept that an activity defined as "sustainable" is also profitable for those who carry it out. It is believed that actions must be carried out for the common good and therefore cannot be monetized: "[...] If we talk about sustainability, a private form is a bit contradictory because it is seen more as a common good" [Focus Group]. However, the first regulations in this regard are emerging, and interest is growing: "[...] the Canton is investing significantly in this area, particularly through the Economic Division." [Interview 4]; "[...] there are various sustainable and ethical projects in Ticino." [Interview 3]; "[...] companies are forced to be responsible and sustainable, especially by public institutions. Thus, they are very interested in initiatives that push towards sustainable practices." [Interview 5].

For this reason, it is essential to have adaptation strategies based on the context (→ Fig. 2).

The initial step should involve integrating the platform as a donation-based, standardized, open-to-all, and free-of-charge. Those investing in the platform do so for philanthropic purposes, which enhances BancaStato's brand image.

The main activities will focus on promotion and marketing to boost community growth and establish partnerships. The platform: "[...] could also provide an overview of local trends and news, promoting sustainability concepts" [Focus Group]. Continuous monitoring and improvement of services, along with the use of contacts and surveys, will allow to adapt to the needs of the target audience and the market situation.

Furthermore, it will be necessary to monitor the situation and make potential improvements.

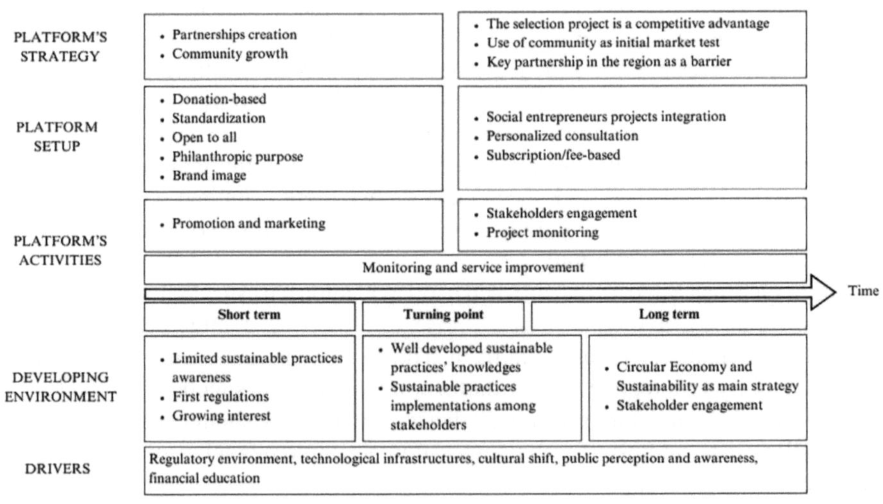

Fig. 2 This figure outlines the adaptation phases of a platform promoting sustainable practices, with key activities, strategy and setup, considering the drivers across short-term, turning point, and long-term horizons

A key aspect to work on during this period is stakeholder engagement. As already mentioned, current sustainability activities are often perceived more as image-building efforts than genuine contributions to society "[…] they do not see the economic return in the short term, as sustainability results tend to be in the long term." [Interview 1]. Therefore, there is a challenge in accepting that a "sustainable" activity can also be profitable "[…] challenges are like those of non-sustainable projects. The fundamental difference is that sustainable projects often emphasize sustainability rather than profit when seeking financing." [Interview 3]; "[…] they do not see the economic return in the short term, as sustainability results tend to be in the long term." [Interview 4]; "[…] the population, is not ready to make a sacrifice solely for the environment" [Interview 5]. Not only is sustainability currently not perceived as something advantageous, but there is also a certain prevalence of competitive thinking within the ecosystem: "[…] There is a lot of competition and little collaboration. If you are small, there are no problems. However, when it is evident that you are building a successful business, others tend to take advantage, especially smaller entities that come later."; "[…] Even in terms of project support, it seems like they don't want to help too much because if they assist you too much, then Zurich or Milan might benefit from it. So, it's more convenient for them if you remain medium-sized, paying taxes here. If the project becomes too successful, they tend to move away". [Focus Group]. This results in silo thinking approach with consequently communication gaps within the system itself: "[…], there is sometimes a lack of communication among the various actors in the Regional Innovation System. This occasionally results in projects bouncing within the system." [Focus Group].

Therefore, an ecosystem emerges in which active collaboration may prove more challenging. For this reason, it is essential to focus on incentives, understanding the necessary activities for each actor, how they can contribute, and the benefits they derive from it: "[…] a good narrative and concrete incentives need to be found." [Interview 5].

This requires, first and foremost, the segmentation of partners. Various examples were presented during the focus group: "[…] regarding the activities of promotion of the projects, institutions might perceive it as a conflict of interest with their initiatives, and it could lead to situations of favoritism that might create political challenges. Therefore, promotion, in this case, should be more about the platform rather than individual projects. […] As for companies, on the other hand, they might more readily support the promotion of projects because it would enable the businesses to establish mutually beneficial collaborations afterward." [Focus Group].

Furthermore, it is necessary to assess how these partners can contribute and what the advantages and disadvantages of these contributions are. For instance, collaboration with local associations or non-profit projects might attract free collaboration from Canton, municipalities, or patriciates, but are considered time-consuming and bureaucratic. Projects involving social entrepreneurs could be more appealing to private companies looking for mid-term returns.

Lastly, incentives need to be defined. It can be provided various forms of revenue that are not exclusively monetary. Revenue streams could include leveraging advertising generated by successful projects or creating a market testing environment to gather feedback before entering the market: "[…] The platform could also act as a method for gathering consensus, so that when interfacing with larger financiers, you have solid foundations." [Interview 3]. These are some of the potential examples identified. Thanks to the cultural development of the regulatory framework, technological infrastructure, public perception and awareness, and financial education, a turning point will eventually be reached where sustainable practices are well developed and widespread, with growing implementation by all stakeholders. The platform has a thriving community and engages stakeholders, with well-known knowledge and implementation in the fields of circular economy and sustainability. In the long term, sustainable practices become an integral part of economies, with widespread and active stakeholder involvement. Therefore, the platform is exploring the possibility of introducing a revenue-based system and offering personalized consultancy services. Active community participation is encouraged through a project selection process that provides competitive advantages.

3.4 Selection Process Optimization

A significant finding of this study, based on literature and competition benchmark is the absence of clear sustainability criteria in current project selection processes in crowdfunding's systems [9, 13–19]. Assessments often overlook these criteria, which could otherwise encourage projects with positive environmental, social, or economic impacts. Integrating sustainability criteria would not only address the growing demand for sustainable initiatives but also help differentiate the platform in the marketplace: "[…] accurate project selection is important to maintain platform credibility." [Interview 1]; "[..] project selection can be a barrier to avoid disappointing proposers with unsuitable projects." [Interview 3]; "[…] This risk could be mitigated through examining incoming projects" [Interview 4].

Moreover, the evaluation process employed by leading platforms often lacks depth, with simplified project submission procedures leading to unrealistic expectations among promoters: "[…] project promoters often have vague ideas about the business economics part of the project." "[…] promoters must actively engage in promotion and attracting funders." [Interview 2]; "[…] they underestimate the effort required for promotion and entrepreneurship despite being guided by programs." [Interview 4]; "[…] they were not familiar with the method and got to know it [a.n.: crowdfunding platform] through word of mouth" [Interview 5]. A notable gap exists in evaluating the entrepreneurial qualities of project promoters. Establishing a systematic framework that assesses both sustainability and entrepreneurial attributes could significantly enhance project filtering and comprehension. At the same time, an overly complex evaluation system would prove burdensome for entrepreneurs: "[…] at the start, when starting an activity, there are many tasks to carry out" [Interview 5]; "[…] the target could be described as people wanting to start their own business, possessing technical knowledge of their field but limited understanding of the administrative and business economics aspects" [Interview 4]. Therefore, it is essential to balance evaluation rigor with ease of use. A framework has been developed to address this (→ Fig. 3).

Data collection, aligned with admission criteria related to sustainability and the Canton of Ticino, alongside promoter profiling and sustainability impact evaluation, enables efficient project assessments tailored to the platform's values. An additional crucial factor is evaluating the promoter's ability to create a necessary fan base. As previously noted, promoters sometimes have a naive understanding of the platform and their role: "[…] they underestimate the effort required for promotion and entrepreneurship despite being guided by programs." [Interview 4]; "[…] they were not familiar with the method and got to know it [a.n.: crowdfunding platform] through

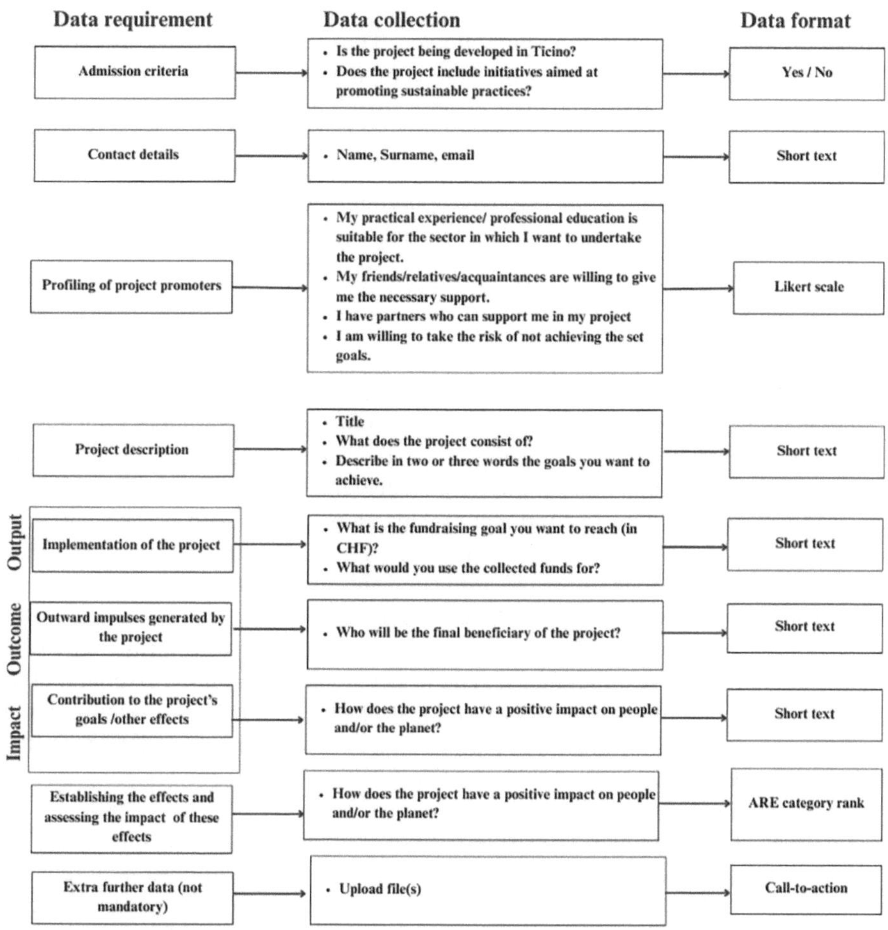

Fig. 3 This figure illustrates a framework for collecting and organizing data requirements for project evaluation. It outlines the data categories, the type of questions or information needed, and the corresponding data formats

word of mouth" [Interview 5]. The awareness of the risk of not achieving predetermined goals is also assessed. Standardized questions allow for a personal evaluation of one's capabilities, offering the platform an initial impression of the individual.

The third and final step involves sustainability impact evaluation. In this case, based on the pilot survey's results, the recommendations of VSOST as guideline [20] were adopted. The first phase entails outlining the project and its intended goals. The second phase allows for a general assessment of the project's impact. Output, outcome, and impact are evaluated through clear and accessible questions. Assessing the effects and measuring the impact proves to be the most challenging part of the selection process, as projects often lack a clear understanding of their business model and, consequently, their impact: "[...] project promoters often have vague ideas about the business economics part of the project." [Interview 2]. To simplify the selection process, promoters will rank the criteria from those they believe will have the greatest impact on those with a more marginal impact. These categories were tested in the market survey and were found to be the most comprehensible for potential users. This enables users to make informed choices, thus avoiding the "my project contributes to all impacts" effect and sparing users from needing to qualitatively evaluate the impact.

On many platforms, users are given the opportunity to upload additional files (e.g., project plans, prototypes, etc.), a feature that has also been included in this platform.

4 Discussion

The proposed sustainable crowdfunding platform in Ticino represents a transformative opportunity to integrate sustainability into the banking sector and enhance local engagement with sustainability initiatives. A stakeholder analysis reveals a diverse range of actors, including institutional organizations, along with private investors and businesses. Each stakeholder contributes unique strengths and varying levels of commitment to sustainability, necessitating the platform to build relationships that effectively bridge these gaps. A bottom-up approach to community involvement is vital, as many local stakeholders perceive sustainability primarily as an image-building exercise rather than a genuine economic opportunity. To counter this perception, the platform should focus on local identity and culture, creating initiatives that resonate with community values. Engaging local voices and demonstrating the tangible benefits of sustainable practices will be crucial for fostering deeper commitment among stakeholders.

However, challenges such as resistance to collaboration and a silo mentality within the entrepreneurial ecosystem must be addressed. The platform must highlight the advantages of inclusiveness and collaboration, presenting a compelling narrative that shows how shared efforts can lead to a greater collective impact. Effective communication strategies will be essential to encourage participation and alleviate concerns about competition. Partnerships will be critical for the platform's success, ensuring credibility and a diverse pool of projects. Clearly defining roles and responsibilities among partners will be vital on avoiding perceptions of favoritism or conflict of interest, particularly with institutions protective of their initiatives.

Integrating the crowdfunding platform into a cantonal bank operation offers both opportunities and drawbacks. The bank's established trust and local presence can enhance the platform's credibility, but a flexible approach is necessary to maintain ongoing community engagement while delivering core banking services. Ensuring user-friendliness and accessibility will encourage broader participation in sustainable funding initiatives. Optimizing the project selection process to include specific sustainability criteria is essential, as current practices often overlook this vital aspect. Establishing a systematic framework for evaluating sustainability impact and the entrepreneurial qualities of projects will help curate a credible portfolio. Leveraging technology to streamline evaluations and enhance user engagement will be crucial for the platform's long-term success. Moreover, measuring the impact of funded projects will build trust and transparency, assuring stakeholders that their contributions yield meaningful outcomes. This transparency will be pivotal in addressing skepticism about sustainability claims, particularly concerns about greenwashing.

In summary, the proposed crowdfunding platform has the potential to significantly advance sustainability efforts in Ticino. By fostering strong partnerships, embracing community engagement, and ensuring a transparent and inclusive project selection process, the platform can effectively tackle the identified challenges. Moving forward, adaptability and continuous dialogue with stakeholders will be essential to align the platform's mission with the evolving landscape of sustainability and community needs, ultimately supporting the local economy and cultivating a culture of sustainability that can serve as a model for other regions.

5 Conclusion

In conclusion, the holistic approach proposed for the sustainable crowdfunding platform in Ticino stands out for its ability to integrate a wide range of actors, promoting active community engagement and valuing local identity. This vision recognizes sustainability not only as an environmental goal but also as an economic opportunity, highlighting the tangible benefits of sustainable practices. Collaboration and inclusivity among stakeholders are essential to overcoming divisions and maximizing collective impact. Additionally, the adoption of a framework for project evaluation that considers both sustainability impact and entrepreneurial qualities contributes to ensuring effective and informed selection. Finally, transparency and impact measurement are vital for building trust within the community, consolidating a lasting commitment to sustainability. This approach not only fosters sustainable progress in the region but also serves as a replicable model for other communities, promoting a culture of responsibility and innovation.

To further advance the development and effectiveness of sustainable crowdfunding platforms, future research should focus on several key areas. First, more detailed studies on the long-term impact of community engagement in sustainability initiatives are necessary, particularly in regions with distinct socio-economic characteristics like Ticino. Additionally, exploring the scalability of such platforms beyond regional contexts would provide valuable insights into their potential to foster sustainability at a broader level.

Lastly, longitudinal studies are needed to assess how the integration of sustainable crowdfunding within banking institutions influences corporate social responsibility strategies over time. This would include the potential for creating synergies between banking operations and community-driven initiatives.

References

1. Johnson, M.P., Schaltegger, S.: Entrepreneurship for Sustainable Development: A Review and Multilevel Causal Mechanism Framework. Entrepreneurship Theory and Practice 44(6), 1141–1173 (2020). https://doi.org/10.1177/1042258719885368.
2. United Nations Brundtland Commission: Report of the World Commission on Environment and Development: Our Common Future. Retrieved from https://sustainabledevelopment.un.org/content/documents/5987our-common-future.pdf (1987).
3. Choi, D.Y., Gray, E.R.: The venture development processes of "sustainable" entrepreneurs. Management Research News 31(8), 558–569 (2008). https://doi.org/10.1108/01409170810892127.
4. Geels, F.W.: Ontologies, socio-technical transitions (to sustainability), and the multi-level perspective. Research Policy 39, 495–510 (2010). https://doi.org/10.1016/j.respol.2010.01.022.
5. Tenner, I., Horisch, J.: Crowdfunding sustainable entrepreneurship: What are the characteristics of crowdfunding investors? Journal of Cleaner Production 290, 1–9 (2021). https://doi.org/10.1016/j.jclepro.2020.125667.
6. Schot, J., Kanger, L., Verbong, G.: The roles of users in shaping transitions to new energy systems. Nature Energy 1(16054), 1–7 (2016). https://doi.org/10.1038/nenergy.2016.54.
7. Edmans, A., Kacperczyk, M.: Sustainable Finance. Review of Finance 1309–1313 (2022). https://doi.org/10.1093/rof/rfac069.
8. State Secretariat for International Financial Matters: Sostenibilità nel settore finanziario. Retrieved from https://www.sif.admin.ch/it/sostenibilita-settore-finanziario (2023).
9. Raiffeisen: eroilocali.ch Condizioni generali di affari. Retrieved from https://www.lokalhelden.ch/ueber/allgemeine-geschaeftsbedingungen_1.html#Pflichten_der_Benutzer (2022).
10. BancaStato: Rapporto Annuale 2022. Retrieved from https://www.bancastato.ch/su-di-noi/cifre-e-pubblicazioni/rapporti-di-esercizio/Bilancio-Sociale-e-Ambientale-2022.html (2022).
11. BancaStato: Bilancio Sociale e Ambientale 2022. Retrieved from https://www.bancastato.ch/su-di-noi/cifre-e-pubblicazioni/rapporti-di-esercizio/Bilancio-Sociale-e-Ambientale-2022.html (2022).
12. BancaStato: Il Mandato pubblico - Origini e significato attuale. Retrieved from https://www.bancastato.ch/DMS/chisiamo/PUBLIC-factsheet_mandatopubblico.pdf (2024).
13. Löher, J.: The interaction of equity crowdfunding platforms and ventures: an analysis of the preselection process. *Entrepreneurial Finance*, pp. 51-74 (2019). https://doi.org/10.1080/13691066.2016.1252510
14. Swiss Financial Market Supervisory Authority: Crowdfunding: Information Sheet. Retrieved from: Swiss Financial Market Supervisory Authority: https://www.finma.ch/it/~/media/finma/dokumente/dokumentencenter/myfinma/faktenblaetter/faktenblatt-crowdfunding.pdf?la=it#:~:text=Il%20diritto%20prudenziale%20svizzero%20non,dalle%20leggi%20sui%20mercati%20finanziari (2020).

15. Progettiamo.ch: progettiamo.ch chi siamo. Retrieved from: https://progettiamo.ch/progettiamo-ch/.(2024)
16. Kick-starter: kick-starter about us. Retrieved from: https://www.kickstarter.com/about?ref=global-footer . (2024)
17. SIG impact: SIG-IMPACT, comment ça marche?. Retrieved from: https://www.sig-impact.ch/ . (2024)
18. Wemakeit: wemakeit chi siamo. Retrieved from: https://wemakeit.com/pages/about. (2024)
19. Triodos bank: triodos bank about us. Retrieved from: https://www.triodoscrowdfunding.co.uk/about (2024)
20. Federal Office of Territorial Development (ARE): Valutazione della sostenibilità: guida pratica per i servizi federali ed altri interessati. Federal Office of Territorial Development (ARE), Department of Environment, Transport, Energy and Communication. Retrieved from https://www.are.admin.ch/are/it/home/sviluppo-sostenibile/valutazione-e-dati/nachhaltigkeitsbeurteilung/nachhaltigkeitsbeurteilung-beim-bund.html (2008).

Situational Analysis of Plastic Waste Management in Pakistan-Policy Perspective and Circular Economy Recommendations

Lala Rukh, Abdul Fatah, and Muhammad Mohsin Memon

Abstract Marine plastic pollution is emerging as a critical global issue, and the river Indus flowing through Pakistan is a significant contributor. About 8–10 million tons of plastic waste make its way to the ocean each year, negatively impacting marine life through ingestion and entanglement. This study is the situation analysis of plastic waste management in Pakistan from a policy perspective. While conducting the situational analysis, globally used, policy instruments are overviewed. The policy instruments used in Pakistan are further discussed in detail based on the recent data and information collected from literature, government reports and stakeholder interviews. A policy inventory is developed, consisting of relevant policies in Pakistan that address plastic or municipal solid waste and its management. Based on a thorough analysis of the Pakistan government's efforts, actions, policies, their implementation and information collected through semi-structured interviews, circular economy recommendations are provided. The circular economy recommendations are developed, keeping Pakistan's current economic state in mind. Given Pakistan's vulnerability to climate change and its energy shortfall, plastic pollution poses an additional environmental challenge for the country. The amazing recycling rate of PET in Pakistan is evidence that it is certainly possible to create a zero-waste model. Pakistan can solve multiple problems by dealing with plastic pollution, such as energy generation from waste. The analysis concludes that adopting a circular economy model presents an opportunity for Pakistan to address plastic pollution while mitigating its energy and environmental challenges.

Keywords Marine plastic pollution · Policy · Public awareness · Situational analysis · Waste management

L. Rukh (✉) · M. M. Memon
University of Galway, Galway, Ireland
e-mail: l.memon1@universityofgalway.ie

A. Fatah
Atlantic Technological University, Galway, Ireland

A. Podleisek et al. (eds.), *Advances in Circular Economy: Bridging Research and Practice*, Springer Proceedings in Earth and Environmental Sciences,
https://doi.org/10.1007/978-3-032-21547-5_2

1 Introduction

Mechanisms are developed at multi-levels; provincial and national. Thus national-level policies or policies developed in other provinces are not concerned with the issue of local-level solid waste management [1]. This disconnect makes policy implementation difficult, weakening the institutional infrastructure and plastic waste to lose the desired attention in the country [1, 2].

Plastic Waste Management (PWM) refers to collecting, transporting, treating, and disposing of plastic materials that have been discarded. Plastic waste management is an essential issue because plastic is a non-biodegradable material, which means it does not break down naturally in the environment. As a result, it can accumulate in landfills and cause environmental problems such as litter and pollution. The PWM addresses human well-being, ecological, aesthetic, limited resource, and monetary concerns related to mismanaged plastic waste [3]. Effective plastic waste management requires a combination of strategies, including reducing the amount of plastic that is produced and consumed, recycling, and properly disposing of plastic waste.

Globally, about 8 to 10 million tons of plastic waste make its way to the ocean each year, negatively impacting marine life through ingestion and entanglement [4]. During 2010–2018, plastic generation as a percentage of total waste generation increased from 12.2 to 13.2, globally [5]. During the COVID-19 pandemic, plastic played a crucial role in protecting us against the virus, and it continues to protect us in the form of protective equipment.

Global plastic production reached about 367 million tonnes in 2020, beginning with a mere 2 million tons of annual plastic production in the 1950s [6]. Every year, global plastic waste generation is more than plastic produced, as post-consumer plastic from previous years also accumulates. Approximately 3% of the world's plastic debris ends in the ocean through multiple waterways, which is fated to be there forever [4]. Plastic is a durable material designed to last long. Further, the ocean environment makes the degradation process even more difficult due to low temperatures and less sunlight.

The river Indus flowing through Pakistan contributes 164,332 tonnes of plastic waste to the sea [7]. Pakistan generates about 6.41 million tons of post-consumer plastic waste in a year, ranking sixth-largest PW producer country. The country's per capita plastic generation rate is 100 g/day, increasing at an alarming rate [8].

The critical limitation of Pakistan's MSW or plastic management and research is data unavailability, and inadequate capacity in terms of human resources, finances, institutions, and technology. Karachi generates about 10,000 to 12,000 tons of waste every day, of which plastic is 20% as per Sindh Solid Waste Management Board (SSWMB). Unfortunately, no proper waste disposal facility is part of the system. As a result, all the waste ends up in the drains and blocks the sewerage lines. The city nearly drained into its filth during the recent rainy season in 2020. Waste mismanagement is a national problem. The vulnerable state of the country's biggest city and commercial hub explains that it is still not taken seriously as a matter of national agenda.

Only in urban areas where proper WMCs (Waste Management Companies) are present, the collection rate is better than the rest of the country [9]. While in 2015, only 57% of urban areas benefitted from collection services, the majority of rural areas have no or minimal systematic collection systems [10]. Due to weak institutional infrastructure and policy disconnection, the municipalities and legal services do not feel obliged to collect, manage, or recycle plastic waste. Thus, it is only at the mercy of scavengers, private junk dealers, and informal sectors.

The informal sector plays a critical role in managing solid waste in the country. Although not part of the national policy or institutional infrastructure the sector is well organised on its individual and small-scale business level. The sector generates about US$1 billion in revenue by collecting, sorting, and recycling dry waste [10]. But only the valuable polymers (e.g., PET), paper, and metal are managed leaving behind low-density PE (LDPE) and other complicated non-recyclable materials (plastics). Although people involved in the informal sector manage to make an income, but at risk of social security and health. As part of this research, we interviewed young boys, but they were very reluctant to come on camera and provide any recorded information. Currently, the government do not claim the energy or material recovery value of any waste, including organic, plastic, glass, metal, or paper. These loopholes signify the importance of capacity building and the need for the Circular Economy (CE) model to be applied in Pakistan's waste management sector.

1.1 Objectives

This study aims to undertake a situational analysis of Pakistan's plastic waste management from a policy perspective and propose circular economy recommendations to save the country from being a significant contributor to marine plastic pollution. In this context, the following are the key objectives of this study:

- Analysis of Pakistan's national-level policies about (plastic) or municipal solid waste management
- Review the national policies and international commitments announced so far.
- Develop policy inventory to document all government efforts since 1997.
- Interview plastic stakeholders to confirm the existing situation.

It is anticipated that achieving these objectives shall help develop circular economy (CE) recommendations based on best practices for effective PWM in Pakistan. Section 2 of this study briefs the methodology applied to achieve these study objectives. Section 3 provides a detailed policy inventory, and Sect. 4 highlights the importance of transitioning to the CE model and provides CE recommendations. Finally, Sect. 5 concludes the study results and analysis.

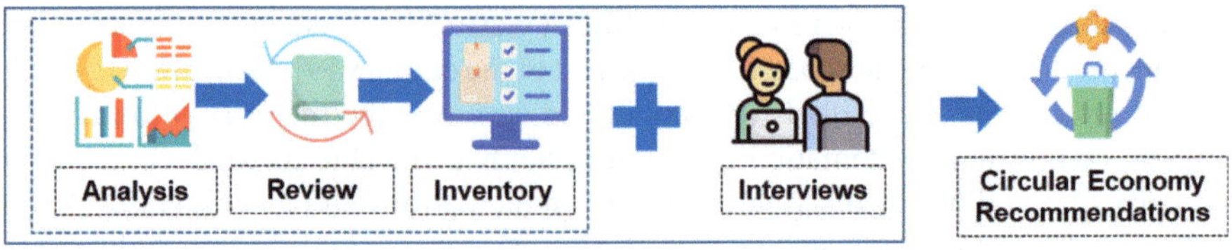

Fig. 1 Methodological framework

2 Methodology

The methodology comprises two parts; the literature review based on the secondary data and stakeholder interview to get the primary data.

2.1 *Step 1: Literature Review (Secondary Data Collection)*

The methodology employed in this article is based on the qualitative data analysis of working papers, national statistics, academic journals, books, newspaper articles, reports and electronic materials. To conduct a situational analysis, it was not possible to rely on scientific literature only as it is very limited special focusing on plastic. Thus, secondary sources such as newspaper articles and reports were also studied. See Fig. 1 illustrates this study's analytical framework.

2.2 *Step 2: Develop of List of Questions and Contacts for Interviews*

A long list of questions was developed to address the identified knowledge gaps after secondary data collection. A broad list of questions was then divided and set according to the six types of stakeholders (i) Government, (ii) Academia, (iii) Plastic Producers, (iv) Plastic Recyclers, (v) Informal sector, and (vi) Consumers involved.

Along list of stakeholders was developed, which was further shortlisted and contacted for the possibility of online interviews during October–November 2021. Finally, 15 stakeholder interviews were conducted, and each interview roughly took 45–60 min.

The interview questionnaire served as a starting point for the conversation, but the main discussion depended on the respondent's answers, as a lot of new information emerged during the interviews. Based on thorough research, circular economy recommendations are provided considering the case of a developing country.

3 Results

Plastic waste management policies are strategies and guidelines that governments, businesses, and other organisations use to reduce the negative impacts of plastic waste on the environment and human health. These policies can take various forms, including regulations, incentives, education campaigns, and infrastructure improvements. Overall, effective plastic waste management policies should be holistic, addressing all aspects of the plastic waste problem and involving various stakeholders, including governments, businesses, and individuals.

Out of 192 countries examined by the United Nations (UN) globally, 127 countries have some form of national policy or address the problem of plastic bags. This significant response is evidence of the world's attention towards the issue of marine plastic pollution. In some countries where national legislation on plastics is not in place, sub-national governments enacted policies or bans to act against plastic bags and other plastic products and successfully received good results [1]. At the same time, it is also noteworthy that some states in the United States (Mississippi and Oklahoma) have implemented pre-emptive policies, i.e., preventing the state from enacting any plastic product ban [2]. Globally, the most common policy is a ban on free retail distribution of plastic bags, followed by 83 countries. Manufacturing and import bans are the second most common regulation, followed by 61 countries. UN examines the distribution of types of federal restrictions in 191 countries [1].

- Fifty-five countries have banned the manufacture, import, and retail distribution of plastic bags
- Nineteen countries have banned the retail distribution of plastic bags
- Ten countries have banned the import and retail distribution of plastic bags
- Seven countries have banned the manufacture and import of plastic bags

Public policy instruments can be divided broadly into three main categories, A: Regulatory, B: Economic, and C: Informative. A: Regulatory instruments can further be subdivided into A1: Affirmative and A2: Prohibitive, and B: Economic instruments can be subdivided into B1: Incentive and B2: Disincentive. The information category can also be subdivided into different categories.

See Fig. 3 represents various policy instruments used globally for PWM. These instruments are discussed elsewhere in detail and thus omitted from this study for conciseness [3, 4]. This study further discusses and categorises Pakistan's policies as per these policy instruments.

To address the rising global concern regarding plastic pollution, the government of Pakistan has recently taken several measures, especially since 2018. Nevertheless, these measures have been significantly impacted by the global pandemic and other administrative and technical constraints to take an implementation phase. Apart from that, presently, the government of Pakistan has set goals to promote waste to energy in the amended Alternate and Renewable Energy Policy 2019 [6], based on the results from the Biomass Resource Assessment Study from Pakistan's 16 significant landfills/dumpsites in the same report.

Pakistan is a developing country yet contributes significantly to global marine plastic pollution through the river Indus. Despite recognising and acknowledging the contribution to the problem, it is only recently that the government of Pakistan has acted and implemented regulations in the form of a prohibition of plastic products and a complete ban on non-biodegradable polythene bags. The pros and cons of introducing a lightweight polyethene bag as an alternative to non-biodegradable bags are yet another debate in the country.

3.1 Policy Inventory

In Pakistan, plastic waste and associated health and environmental concerns are identified, and thus the government has taken several actions, evident from the following policy inventory. Although 100% implementation of the policies or bans on plastic bags has mainly not achieved, the government has kept revising and updating the bans and national waste management strategies. See Fig. 2 represents the progress of Pakistan's federal and provincial policies and bans from 1997 until 2021.

See Fig. 2 briefly summarises how bans and guidelines have been repeatedly imposed and revised due to implementation failure. Following are these policies reviewed and analysed in detail. Pakistan has enacted various national policies and regulations aimed at managing plastic waste and promoting environmental sustainability. These include:

- Pakistan Environment Protection Act (PEPA) 1997: Establishes the Pakistan Environmental Protection Agency (EPA) and sets standards for air, water, and

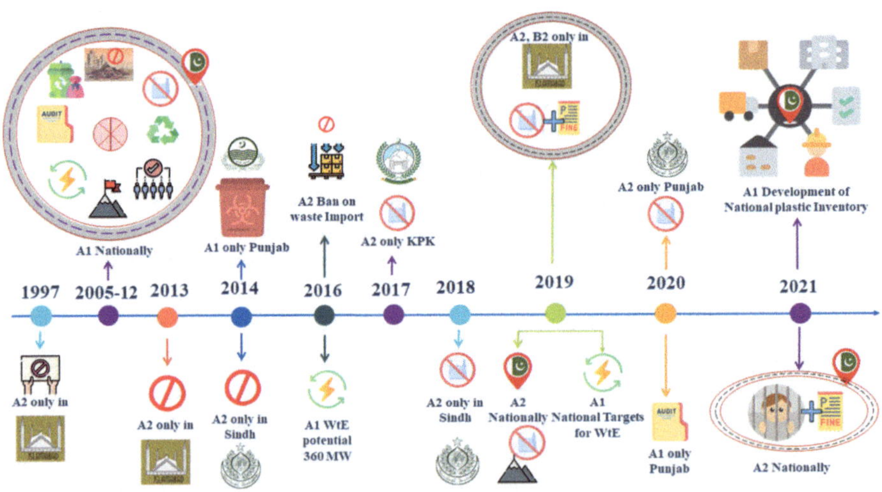

Fig. 2 Policy progress in Pakistan 1997–2021

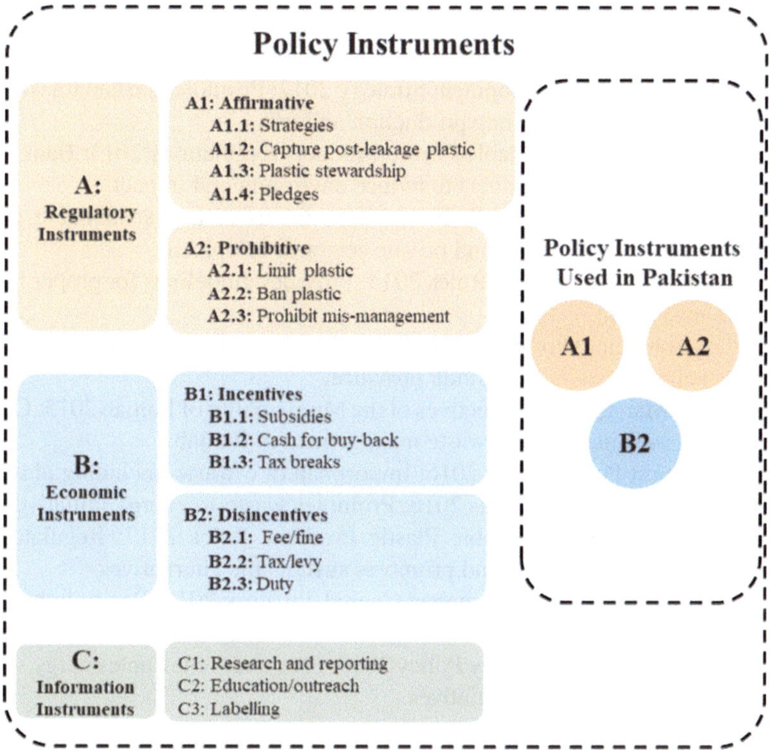

Fig. 3 PWM policy instruments observed globally

land quality. It bans non-biodegradable plastic products in the Islamabad Capital Territory [7].

- National Environment Action Plan (NEAP) 2001: Identifies environmental issues and strategies for addressing them, promoting integration of environmental considerations into all sectors [8].
- National Environment Policy 2005: Sets principles like polluter pays, precautionary principle, intergenerational equity, and public participation. Aims at improved environmental quality, sustainable development, climate change mitigation, and biodiversity conservation [9].
- Guidelines for Solid Waste Management 2005: Provides recommendations for safe and efficient solid waste management, including waste reduction, reuse, recycling, and proper disposal.
- Pakistan National Sanitation Policy 2006: Aims to improve sanitation and hygiene, reduce inadequate sanitation's adverse impacts, and ensure access to safe sanitation services.
- Clean Development Mechanism (CDM) National Operation Strategy 2006: Encourages GHG reduction projects contributing to sustainable development.

- National Climate Change Policy 2012: Addresses climate change impacts, including regulation of waste management practices [10].
- National Sustainable Development Strategy 2012: Promotes sustainable development, discouraging plastic bag production and use.
- Prohibition of Non-Degradable Plastic Products Regulations 2013: Bans certain non-degradable plastic products to reduce environmental impact.
- 11th Five-Year Plan 2013–18: Emphasizes inclusive and sustainable growth, investment in infrastructure, and private sector development.
- Hospital Waste Management Rules 2014: Provides guidelines for proper hospital waste management.
- Sindh Environmental Protection Act 2014: Regulates environmental activities and protects natural resources in Sindh province.
- Solid Waste Management Directives of the Municipality of Punjab 2015: Outlines policies for sustainable solid waste management in Punjab.
- Pakistan Import Policy Order 2016: Import ban of e-waste, including plastics.
- Final Report on Biomass Atlas 2016: Promotes waste-to-energy initiatives.
- Prohibition on Non-Degradable Plastic Products Rules 2017: Regulates non-degradable plastic products and promotes sustainable alternatives.
- Ban on Polythene Bags in Islamabad Capital Territory 2019: Ban polythene bags to reduce plastic pollution.
- Alternate & Renewable Energy Policy 2019: Promotes renewable energy sources, including waste-to-energy initiatives.

Although Pakistan is also a signatory of international commitments to regulate plastic and other waste, there is a gap in the knowledge confirming the ground realities of the obligations and agreements that are being followed. From the national policies and plans analysis, it is evident that severe actions against plastic are very recent, targeting only plastic bags and region limited. Pakistan has ratified international agreements like the London Convention, MARPOL, Stockholm Convention, Basel Convention, and Rotterdam Convention to address marine pollution and hazardous waste management. These policies and regulations reflect Pakistan's commitment to sustainable development and environmental conservation, aiming to mitigate the adverse impacts of plastic waste and promote a cleaner, healthier environment.

4 Need for Transition to Circular Economy Model

Circular economy models are designed to promote the sustainable use of resources, including plastic, by reducing waste and promoting the reuse, repair, and recycling of materials. Implementing a circular economy model for plastic waste management could help address this issue by reducing the amount of plastic produced and discarded and promoting responsible use and management of plastic resources. This could have several benefits for Pakistan, including reduced environmental pollution, improved public health, and economic benefits through creating new jobs and businesses in the recycling and reuse sectors. There are several potential benefits to implementing a circular economy model for plastic waste management: environmental, economic, social benefits and resource conservation.

The plastic industry in Pakistan progressed throughout the country like a fire in a jungle. Without much progress in the recycling side of the business. This may be attributed to the fact that most of the collection, sorting, and sale of plastics and all other recyclables is mainly done only by the private sector and junk dealers without much government involvement.

To implement a circular economy model in the waste management sector. Pakistan government shall have to take a serious interest in this matter. As CE is not only achieved by achieving 3R (Reduce, Reuse, and Recycle), it requires redesigning and rethinking the entire material flow from the cradle to the grave. Pakistan can use national resources and international support from CPEC (China-Pakistan Economic Corridor) to design the country's plastic production, consumption, and recycling. The increasing amount of non-degradable plastic in MSW composition ending in landfills needs immediate government attention. The informal or private business sector cannot alone solve this problem. The waste can be turned into a resource by recovering its material or energy value, provided adequate policies, implementation, and regulations.

In other cases, following this linear economy model in a developing country, a penalty/ban/fine imposed on a single plastic product does not guarantee effective PWM. These can only help raise awareness and seek public attention for the time being. An ambiguous ban on plastic bags has only created a state of confusion in public rather than bringing any positive results. Multiple and repeated ordinances by the federal and provincial governments also created confusion in the country, especially as waste management comes under only provincial authority after the 18th amendment in the constitution of Pakistan. The waste management sector has been going through constant institutional restrictions over time, which can be another reason for delays and effective management in the country.

The policy inventory shows that policies, rules, and guidelines, including regulations for import/export and hazardous waste, are already present in the country. However, the long-term and continuous implementation of the same has not been observed so far. Thus, there is an immediate need to revisit the bottlenecks of the policy implications.

It is also pertinent to mention that oxo-biodegradable lightweight plastic bags made their place on the market immediately after the ban on non-biodegradable plastic bags. Thus, the public had little opportunity to understand the prohibition and its replacement. On the one hand, the government considers they have played its role and found the solution, whereas the public understands that the ban is not in place, and nothing has changed in the country before and after the ban. Pakistan also exempts the ban on plastic bags used for sanitation or waste storage and disposal [11].

Pakistan is still far behind in terms of policy implications as compared to other neighbouring countries. Before expecting any policy results, associated institutions must be strengthened so that the national and provincial governments stand on one page while making decisions.

4.1 *Identification of Circular Economy Opportunities Using Policy Interventions*

The top-down and bottom-up approaches, ranging from segregation at source to public awareness, must be enacted at local, provincial, and national levels to realise the CE model for PWM in the country. Pakistan's annual per capita contribution to plastic marine debris is about (0.73–1) kg, compared to Sri Lanka (1.5–3 kg), Philippines (2.6–7 kg), Vietnam (2.9–7.7 kg), Egypt (1.5–4.2 kg), and Nigeria (0.6–1.7 kg) [12].

Following realistic CE recommendations are provided based on the country's struggling situation. The achievement of these shall shift the direction of PWM practices, i.e., capacity building, infrastructure improvement, and institutional restructuring towards a more functioning ecosystem.

- The roles and responsibilities of stakeholders and government organisations must be clearly defined.
- The opportunity for time and alternatives must be provided to the public when a particular product is banned.
- Source-based waste segregation and the involvement of information technology can help solve the problem.
- There is an existing infrastructure for plastic recycling, as almost 99% of PET is recycled in Pakistan. But only through the private sector with the help of junk dealers. If the government steps in, a similar process and chain can be utilised for recovering the material and energy value of plastics, among other recyclables. Private recyclers are helping the government manage the waste; it is recommended to provide subsidies to promote the business in the country.
- One way of expressing interest can be capacity building of these junk dealers, and private recyclers to make the collection and sorting process easy.
- Well-segregated plastic waste can be further utilised for electricity generation in Pakistan. World Bank has already conducted a quantitative analysis of 16 significant landfills in the country.
- It is recommended to expand the policy focus beyond plastic bags alone. Plastic sachets, for example, are the most trouble-causing form of plastic waste. Immediate policies and a ban must be put on the same.
- Public participation is the key to PWM. From the pilot models running in Islamabad, Karachi, and Lahore, it is evident that the public is willing to participate, segregate the waste, and pay for basic service charges. In Islamabad, effective awareness campaigns and seminars are regularly arranged, which are then covered

by local news channels and media personnel, in addition to strict monitoring of the plastic bags ban, as severe fines/punishments can be given to violators. Using the Islamabad role model, public participation can be sought throughout the country. For which provincial authorities shall have to participate actively.

5 Conclusion

This study provided a detailed review of the status of national and provincial-level policies, bans, guidelines, and strategies announced so far. In addition to the international commitments that Pakistan is a signatory country. Based on this review, the key objectives of this study were achieved, introducing globally used policy instruments for plastic waste management in Sect. 3, and providing policy inventory for Pakistan with further discussion of all Pakistan's policies in Sect. 3.1. Stakeholder (semi-structured) interviews were a very interesting aspect of this study, which broadened the scope and confirmed or corrected the secondary data collected from the literature review.

Based on a comparative study, i.e., a literature review of policies, and stakeholder interviews, it is concluded that plastic waste management in Pakistan is not very effective. The government only puts efforts into collecting mixed municipal solid waste, which is then dumped into non-sanitary landfills. The only kind of plastic that has sought the government's attention, as reflected in policy documents, is 'the plastic bag'. Despite the Pakistan government's repetitive bans on the production, consumption, and sale of plastic bags, it is still a very common product in the country. The fact is blanketed by rebranding the plastic bags as oxo-biodegradable plastic bags. It is pertinent to mention here that the reuse and recycling of these so-called oxo-biodegradable plastic bags is an even more complex phenomenon than ordinary plastic bags.

Nevertheless, plastic bags are not the only elephant in the room here. For effective PWM, the country needs to redesign the entire economy from linear to circular. Therefore, this study provides suitable circular economy recommendations for a developing country.

It was realised that the primary literature does not provide sufficient up-to-date information about the current situation of PWM in Pakistan. Thus, this situational analysis is conducted from different aspects, and policy, as represented in this study, is one of them. The situational analysis is based on primary and secondary data collection and stakeholder interviews. Stakeholders are from the business, government, academia, public, and private sectors.

References

1. UN, "Legal Limits on Single-Use Plastics and Microplastics: A Global Review of National Laws and Regulations," 2019.
2. NCSL, "State Plastic Bag Legislation," National Conference of State Legislatures.
3. J. K. Abbott and U. R. Sumaila, "Reducing marine plastic pollution: policy insights from economics," *Rev Environ Econ Policy*, vol. 13, no. 2, pp. 327–336, 2019.
4. G. Finnveden *et al.*, "Developing and evaluating new policy instruments for sustainable waste management," *International journal of environment and sustainable development*, vol. 11, no. 1, pp. 19–31, 2012.
5. GoP, "Economic Survey of Pakistan 2019-20," 2020. [Online]. Available: http://www.finance.gov.pk/survey/chapters_17/Pakistan_ES_2016_17_pdf.pdf
6. Government of Pakistan, "Alternative and renewable Energy Policy, 2019," 2019.
7. GoP, "Pakistan Environment Protection Act (PEPA) 1997 of Pakistan," 1997. Accessed: Jan. 05, 2023. [Online]. Available: https://www.elaw.org/system/files/Law-PEPA-1997.pdf
8. Ministry of Forest & Environment, "National Environment Action Plan (NEAP)," 2001. Accessed: Jan. 06, 2023. [Online]. Available: https://www.thegef.org/sites/default/files/ncsa-documents/National_Environment_%26_Health_Action_Plan_cover_merged.pdf
9. GoP, "National Environment Policy 2005," 2005. Accessed: Jan. 05, 2023. [Online]. Available: https://pbit.punjab.gov.pk/system/files/National%20Environmental%20Policy%202005.pdf
10. GoP, "National Climate Change Policy," 2012, *Islamabad*.
11. Mansoor A, "Environment: how plastic is killing us," 2019. Accessed: Dec. 07, 2021. [Online]. Available: https://www.dawn.com/news/1477373
12. J. R. Jambeck *et al.*, "Plastic waste inputs from land into the ocean," *Science (1979)*, vol. 347, no. 6223, pp. 768–771, 2015.

Situational Assessment of Plastic Waste Management in Pakistan: Insights from Public Awareness and Stakeholder Engagement

Lala Rukh, Abdul Fatah, and Muhammad Mohsin Memon

Abstract Plastic pollution is becoming a rising challenge for Pakistan, with river Indus being one of the most (plastic litter) polluting rivers in the world. Bans on disposable plastics have been largely ineffective, and a lack of awareness and operational recycling facilities have exacerbated the issue. This study has assessed the current state of plastic waste management (PWM) in Pakistan, focusing on public awareness tools and stakeholder engagement in the country. The key stakeholders are identified as (i) the Government, (ii) Academia, (iii) Plastic Producers, (iv) Plastic Recyclers, (v) the Informal sector, and (vi) Consumers. Through semi-structured interviews and a power-interest analysis, this study has identified the roles and challenges of these stakeholders in managing plastic waste. Also, a change to the power-interest map is recommended. The recommended power-interest map emphasizes the critical role of the Pakistan government (provincial and national level), both in capacity building and policy enforcement, to support other stakeholders in their efforts toward an effective plastic waste management system. Finally, the need for increased public awareness, corporate responsibility, and the transition toward a circular economy is highlighted to address Pakistan's plastic waste crisis.

Keywords Action · Education · Plastic abatement · Plastic pollution · Public awareness · Power-interest map · Stakeholder engagement analysis

---

L. Rukh (✉) · M. M. Memon
University of Galway, Galway, Republic of Ireland
e-mail: l.memon1@universityofgalway.ie

A. Fatah
Atlantic Technological University, Galway, Republic of Ireland

A. Podleisek et al. (eds.), *Advances in Circular Economy: Bridging Research and Practice*, Springer Proceedings in Earth and Environmental Sciences,
https://doi.org/10.1007/978-3-032-21547-5_3

1 Introduction

Plastic pollution is yet another rising challenge in Pakistan. The River Indus flowing through Pakistan is the second most (plastic) polluted river in the world [1]. In 2018, the River Indus ranked second most plastic-polluting river in global oceans, carrying more than 90% of waste, which is plastic [2]. Several bans on disposable plastics have been ineffective in Pakistan [3]. For example, Sindh, one of the country's provinces, has repeatedly failed to implement a ban on plastic bags since 2006, following up in 2014 and then again in 2019 [3]. The same is true of the federal government, as prohibiting non-biodegradable plastic bags in 2009 was a significant failure.

Although public awareness in the country about environmental issues including plastic waste management has been rising over the past decade, it is still insufficient. The lack of awareness on one hand and little to no operational plastic recycling facilities on the other are the major drivers behind the existing catastrophic situation of plastic waste management in the country [2]. If the business-as-usual scenario continues, the greenhouse gas emissions from plastic could reach 1.34 billion tons per year [2].

Unfortunately, plastic bags are the only plastic product that has caught the government's attention, as evidenced by several policy measures to either prohibit its production/manufacturing and sale or ban its consumption. Well, it is also true that plastic bags are one of the most common wastes found in general garbage and that ends up on beaches along Pakistan's coast [2]. Pakistan's 2019 plastic bag ban halted the production of 600,000 kilograms per annum [4]. A post-ban survey carried out in September 2020 showed that 80% of the respondents felt that the ban played a major role in increasing their knowledge of plastic pollution (ibid.).

As per a risk modelling firm RMS, the Pakistan floods of 2022 were the tenth most expensive climate disaster over the last decade [5]. It is relevant and important because the country's economic management has been falling from one crisis to another. Poor public finance management has caused macroeconomic instability, high inflation, poor public services, criminal neglect of the social sectors, widespread corruption, crippling power outages, growing unemployment, deepening poverty and a deteriorating debt profile [6]. In such a state, Pakistan is not able to face the consequences of yet another rising challenge of (marine) plastic pollution. At the same time, it is equally valid that in such difficult economic situations, marine plastic pollution might be the least of the worries for the public and government.

Thus, this study aims to understand the existing situation of Plastic Waste Management (PWM) from a public awareness perspective and analyse stakeholder engagement. The stakeholders are identified as (i) the Government, (ii) Academia, (iii) Plastic Producers, (iv) Plastic Recyclers, (v) the Informal sector, and (vi) Consumers. However, NGOs, INGOs, and international development sectors can be a separate stakeholder category for this study; they are included in the government sector. All stakeholders are understood as plastic consumers by default. The following objectives are set to achieve the study objectives systematically.

The objectives of the study encompass several key aims: firstly, to ascertain the existing Public Awareness tools utilized for effective Plastic Waste Management (PWM) in Pakistan and assess their current implementation status; secondly, to pinpoint and conduct interviews with pertinent stakeholders within the PWM sector to gain insights into their perspectives and practices; and finally, to construct a Power-Interest map, delineating the level of influence and interest of various stakeholders in PWM, thereby facilitating a comprehensive Stakeholder Engagement Analysis.

The following sections of this study provide a methodology description used to achieve the objectives and later shed light on the results of the study in terms of public awareness, public awareness tools for awareness about plastic pollution, notes from semi-structured interviews conducted, and finally, stakeholder analysis. Subsequent paragraphs, however, are indented.

2 Methodology

This study is divided into three sections based on three objectives as identified above. The methodology applied for each one is briefed as follows. First, to understand the public awareness tools about PWM in the country, their application, and identifying stakeholders. The literature review allows analysis of the current situation of PWM in Pakistan from the public awareness perspective. It is pertinent to mention here that this study has equally considered literature scientific and grey literature, including publicly accessible documents, student papers, national and international reports, and especially news from reliable resources.

A literature review from diverse resources helped to identify the relevant and multiple stakeholders. The knowledge gaps from the review were used to develop a long list of questions. A long list of three hundred stakeholders was created initially, which was categorised as (i) Government, (ii) Academia, (iii) Plastic Producers, (iv) Plastic Recyclers, (v) Informal sector, and (vi) Consumers. Further, almost fifty short-listed key stakeholders were approached for a semi-structured interview. Following responses and scheduling, fifteen semi-structured interviews were conducted.

While performing the qualitative research study, mainly three interview types can be identified: Structured type, Semi-structured, and Narrative [1]. Based on this information, the semi-structured interviews were conducted prominently to obtain primary data from experts in the PWM sector. The semi-structured interview type has a focus, but the questions are flexible depending on the direction of the participants' replies. Interviewers prepare predetermined open-ended questions, but other questions create a conversation between interviewees and interviewers.

Six different questionnaires designed according to stakeholders were developed. These were further focused on specific objectives concerning the stakeholders involved to keep the interviews short and efficient. These interviews were conducted in November 2021 over an online call and (audio) recorded with the permission of the interviewees. As a token of appreciation for the interviewees' time, a donation of 500 PKR for each interview was made to Shaukat Khanum Memorial Cancer Hospital & Research Centre in the name of the interviewees.

Finally, a broader and more open discussion during these interviews significantly helped to develop the Power-Interest map for stakeholder engagement analysis. After the interview process, much of the information was corrected, updated, or added to the literature.

3 Results

The adverse impacts of plastics are poorly understood and, more importantly, not disseminated among the common people, who are the everyday consumers of this pervasive commodity. The plastic industry in Pakistan is expanding in food packaging and other major industries [2]. New plastics such as Fiber-Reinforced Plastic (FRP) further add to the complexity of the problem. Their recyclability is even more challenging than ordinary plastic. The nationwide ban on non-biodegradable bags in Pakistan is having a rising impact on consumption patterns. Also, the consumer tends to understand the importance of using alternate bags as a healthy option for the environment after realising the non-degradability of plastics and their associated effects on nature and humans [3]. More people, organisations, and producers in the developed world understand the concept of individual responsibility towards the environment. However, the story of developing countries is exceptionally different [4]. Most of the population still ends up littering the cities, despite the provision of waste collection points.

Note from Interview # 11: Karachi residents treat Karachi as a temporary home. There is an influx of people from all around the country searching for work, but no one owns the city, not even the city government. This is the main difference between Islamabad, Karachi, and Lahore regarding policy implementation. Plastic is the least of the city's troubles, among several other mishaps happening every day. Among fifteen interviews conducted, it was found that plastic pollution is causing damage to multiple resources including agriculture and fish industry along the river Indus. The same agriculture and fish industries contribute significantly to plastic pollution. Raw plastic is widely used in various industries and PWM is not the priority of these industries, as PWM requires additional expenses. New kinds of plastics are adding to the complexity of the issue, instead of solving it. Pakistan does not have enough technical and capital capacity to deal with ordinary or Oxo-biodegradable (OBD) plastic bags.

Globally, various public awareness tools are considered for raising public awareness about the plastic pollution issue, as shown in Table 1. They are divided into two categories, i.e., A: Action and B: Education. Action tools are defined as tools that, along with raising awareness, have some positive impact on reducing plastic waste, even at a minor level. In contrast, Education tools are defined as the tools used only for education and raising awareness. The application of these tools involves

Table 1 Public awareness tools used globally

A: Action tools	B: Education tools	
A1: Ocean clean-ups	B1: Plastic footprint	B9: Television ads
A2: Plastic-free day/week/month	B2: School curriculum	B10: Radio shows
A3: MARLISCO	B3: Demonstration project	B11: Mobile apps
A4: Using art as a tool	B4: Plastic disclosure project	B12: Webinars
A5: Promoting 3R concept	B5: MARLISCO	B13: Hackathons
A6: Waste segregation efforts	B6: Taking the help of art	B14: Social media influencers
A7: Writing to authorities	B7: Essay competition	B15: Rallies and march
A8: Mobile apps	B8: Brand audit	B16: Community meetups

all major stakeholders, (i) Government, (ii) Academia, (iii) Plastic Producers, (iv) Plastic Recyclers, (v) the Informal Sector, and (vi) Consumers.

3.1 Public Awareness (PA) Tools Used in Pakistan

A post-plastic bag ban survey carried out in September 2020 showed that 80% of the respondents felt that the ban played a major role in increasing their knowledge of plastic pollution [5]. The survey also showed that all the retailers interviewed had shifted towards using alternative bags (ibid.). The following examples of PA action and educational tools are used in Pakistan.

Action Tools

Action Tools are the tools that, along with raising awareness, have some positive impact in reducing plastic waste, even at a minor level; some of the example applications in Pakistan are as follows.

Ocean clean-ups in Pakistan. The activities involved in the ocean clean-up and beach clean-up activities are action-based public awareness tools. When hundreds of people are gathered for cleaning activities, they engage the local community by disseminating the knowledge and coming into action for a global cause [6].

In Pakistan, in April 2017, Khushhali Microfinance Bank, in partnership with WWF-Pakistan, conducted a beach cleaning and mangrove plantation activity. Awareness about plastic pollution and marine litter by the WWF Climate Team followed the drive. In April 2019, 800 participants from diverse backgrounds joined for coastal clean-up in Karachi. During the coastal clean-up drive, 140 bags of waste were collected from the 6 km beach line [7]. Note from Interview # 1: The Beach Clean-up Karachi 2021 project aimed to engage volunteers to help manage trash at Seaview Beach in Karachi. The collected garbage will be measured, sorted, and sent to recycling centres. Non-recyclable trash is planned to be used for making art pieces.

Celebrating plastic-free day/week/month. Beginning with good habits, one can change lifestyle. Campaigns such as "Plastic Free July" can help eliminate or minimise single-use plastic waste from our lives and help us get more aware of the agenda, thereby taking the challenge. Mostly, these interesting challenges do not end after one month; they continue to the next level [8].

Note from Interview # 7: The Plastic Free Pakistan team initiated the practice of cleaning while hiking in March 2020 and named 'Let's Clean up our Act at Margalla Hills.' This activity was performed in collaboration with the Islamabad Wildlife Management Board and Ecological. This is one of many practices the Plastic Free Pakistan team regularly conducts to showcase how we can contribute as a community. Hunza district in Pakistan set an example by becoming the first district in Asia to be plastic-bag-free in April 2019. Ten thousand cloth bags among the people were distributed after the formal imposition ceremony by the city government.

MARLISCO: MARine Litter in Europe Seas: Social AwarenesS and CO-Responsibility MARLISCO has contributed to setting a clear picture regarding Marine litter in each of the European Regional Seas as well as to towards a better understanding of the science-society nexus.

Taking the Help of Art to Raise Awareness About Plastic Pollution. Several forms of art can be helpful tools to raise awareness about plastic pollution and divert some of the plastic litter from landfills [9]. **Note** from Interview # 15: Pakistan's Green Earth Recycling company makes attractive outdoor furniture, play areas, gazebos, and desks from recycled plastic waste. The company aims to raise environmental awareness and make it economically viable.

Promoting the concept of 3Rs. The idea of sustainable circular economy (CE) is undoubtedly much broader than 3R, but 3R is an essential element of the whole CE model [10]. **Note** from Interview # 5: The government of Pakistan is introducing the concept into all new policies and plans to ensure that the public and private sector follows the reduce-reuse-recycle pattern.

Waste Segregation Efforts. Although segregated waste is still not adequately managed in Pakistan. **Note** from Interview # 7: When people notice the multiple bins, they consider the importance of recycling and material recovery, from their personal experience.

Education Tools

Education Tools are the tools used only for education and raising awareness. These tools involve all major stakeholders: government, academia, the private sector, the informal sector, consumers, and recyclers. Education tools are further classified as follows.

Plastic Footprint. The plastic produced decades back is still in the environment emitting carbon and other GHG emissions. The concept of plastic footprint helps one identify and realise the amount of plastic and associated carbon released into the atmosphere by continuing the current lifestyle [11]. A plastic footprint is an individual tool for self-realisation and a general understanding of how much plastic we leave behind in our everyday consumption.

School Curriculum. Fostering responsible behaviours from an early age can effectively address issues like plastic pollution. Educating students on these matters early on leads to better, long-term outcomes compared to imposing forced behaviours. Recently the Government of Pakistan applied a Single National Curriculum (SNC) in different phases. The first phase of SNC is being used at the primary level from April 2021 [12]. As part of the curriculum revision, recent information such as COVID-19, climate change, and plastic pollution has been introduced in the syllabus, which was not part of the previous syllabus (ibid.).

Demonstration Projects. However, the national curriculum has been recently revised and includes the necessary information about significant societal challenges such as plastic pollution (ibid.). Still, demonstration projects as part of the school curriculum or super curricular activities are very effective in raising awareness and training young mindsets to become responsible citizens. For example, Science Fuse-Pakistan delivers hands-on methods and experiments to understand science as part of their summer camps and fairs.

Taking the help of art. Social awareness about plastic pollution can be raised through almost all art forms, photography, performance, events, music, poetry, illustrations, and infographics [9]. These can be further broadcast on national television and radio channels, creating a significant impact. Animations can directly target the young audience. **Note** from Interview # 8: For example, Commander Safeguard and Dettol Warriors are famous mascots and role models among Pakistani children. Although used for advertising the soaps of brands, children learn exciting lessons about the importance of cleanliness.

Essay Competition among Youth to measure awareness. Teachers can assess the understanding of youth towards the issue through essay and video competitions. Students' submissions reflect their experiences with the PWM in the country. Independent online education organisations have recently started to arrange essay competitions and video contests to raise awareness about plastic pollution and other related social problems. These competitions are very efficient as young students will get a chance to research and learn more about social issues in detail.

Brand audit. Brand Audits are an essential awareness tool to recognise the most polluting entities and hold them accountable for their contribution to the problem [13].

Television Ads. **Note** from Interview # 10: The pandemic time is extraordinary evidence of how much television ads can help bring up public response towards social issues and action against them. For example, cute little animations of coronavirus aired on national television alerted even very young children as young as 2–3 years old. Even kids realised after attractive and animated television ads what coronavirus is and how to deal with that. Similarly, short TV commercials must also highlight plastic pollution to bring public awareness.

Radio Shows or Podcasts. Likewise, radio shows and podcasts are other platforms to raise awareness about plastic pollution. The message sent in the regional language of people with typical examples of negative impacts further highlights the issue.

Mobile Apps and YouTube (Ads). App and YouTube ads are effective as the consumer is forced to watch the 5 to 15-second ad that leaves a mark on the consumer's mind. The government must invest in such tools instead of putting a blanket ban and expecting results.

Webinars/In-person Seminars. As the curriculum cannot be revised instantaneously, webinars effectively communicate the latest research and information among youth, students, and the public [14]. The College of Earth and Environmental Sciences, University of Punjab, Lahore, has been conducting a series of webinars and highlighting the agenda of 'Plastic Pollution Free Pakistan' during July 2021. NUST School of Chemical & Materials Engineering (SCME) and NUST Office of Sustainability actively engage with national government and international development bodies such as PSF, UNDP, UNESCO, and Coca-Cola Foundation to conduct seminars for students' awareness regarding plastic waste and the importance of recycling.

Hackathons (Sciathons) are a new trend of solving a complex problem with limited resources in a limited time frame. Recently, a Pakistan case study, "Let's Fix the Marine (Non-Recyclable) Plastic Litter Issue!" was part of an international Sciathon (Competition) [15]. The project aimed to provide solutions to manage non-recyclable plastic through innovative and new methods that do not harm the environment. Such programs are also a great source of raising awareness about the problem of plastic pollution.

Social Media Influencers and Celebrity Endorsement. Social media influencers and celebrities influence the public; they can take the lead and help achieve results. Local celebrities and influencers urge Pakistanis to step up, and (#FaceThePlastic) is a slogan by Unilever Pakistan to raise plastic awareness in the country [16]. During the pandemic, celebrities and influencers noticed a rapid growth in the use of plastic. Public figures like Sarwat Gilani, Shaneira Akram, and Muniba Mazari took to the internet to share a video about how much waste we produce in Pakistan and urged the public to face plastic. Shaneira Akram shared some alarming statistics on Twitter about Pakistan's plastic pollution. She also raised awareness about toxic medical waste on Clifton Beach, Karachi.

In collaboration with Hoga Saaf Pakistan, WWF-Pakistan arranged a webinar titled 'Beat Plastic Pollution' to raise awareness about plastic pollution. Media personalities and officials, including Elin Burns, Ayesha Omar, Frieha Altaf, Momal Shaikh, Yasir Hussain, Hareem Farooq, Sharmeen Abbas, and Tooba Siddiqui, joined the webinar [17]. Local celebrities generally influence the local public in the country and tend to follow their guidelines or what they do. In some cases, it may also be their only source of information.

Rallies and March. **Note** from Interview # 12: Protests, Marches, and Rallies are also common sources of bringing the public's attention to issues such as plastic pollution. It immediately catches public attention and makes it to a newspaper cover. Thus, information and messages reach more people effectively.

Community Meetups. **Note** from Interview # 12: Community meetups in mosques and other community halls and discussions can bring positive results, a general practice globally. This subject can be included in Friday's sermon, as people listen to that very delicately. As confirmed during stakeholder interviews, the issue is only discussed under the umbrella of 'cleanliness is the half of faith.'

In Pakistan, the most effective tools are social media and television, as these are the two platforms accessible to most of the public. As described in the examples above, when social media celebrities take notice of social problems and raise their voice, even the government act instantaneously. During stakeholder interviews, almost all the interviewees also acknowledged that social media is the best tool for the public of Pakistan.

3.2 *Stakeholder Analysis*

Stakeholder analysis for plastic waste management in Pakistan involves identifying and evaluating the interests, influence, and potential impact of key groups and individuals who are affected by or have an impact on the management of plastic waste in the country. The goals and motivations of each of these stakeholders may vary, and it is crucial to understand their perspectives to engage them effectively in plastic waste management efforts. Analysing their interests, resources, and potential for collaboration or resistance can help develop effective strategies for addressing plastic waste in Pakistan.

Nevertheless, other vital factors are also required in addition to stakeholder roles and responsibilities. The most important of all is the time. It takes time to change habits and shift from using a widely available disposable option to practising carrying a reusable item. Thus, the opportunity for time and alternatives must be provided before imposing an outright ban.

From the literature review and stakeholder interviews, it is realised that some public of Pakistan are willing to pay fees, participate, segregate the waste, and convert it to reusable bags. Also evident from the Islamabad city example. Strict policy measures, such as fines and punishment, further ensure the implementation of government-introduced measures. Nevertheless, it is high time to expand these measures beyond Islamabad territory. Education plays a crucial role in participation, as the public in Islamabad's twin city Rawalpindi behaves differently, and plastic bags are as standard as always. Unfortunately, awareness is often only for discussion without results; for example, Nestle ironically branded a paper straw wrapped in plastic packaging. Nevertheless, this brings public attention to the matter under discussion. Pakistani public needs strong long-term measures for awareness and policy implementation tools.

The key stakeholders considered for this research are 1-Government, 2-Academia, 3-Plastic Producers, 4-Plastic Recyclers, 5-Informal Sector, and 6-Consumers (each stakeholder is a consumer also). Thus, every stakeholder's understanding of national policy, bans, and waste management model is worth the analysis. (see Fig. 1), which represents an example of stakeholder engagement in the above-mentioned awareness action and educational tools for public awareness about plastic pollution in Pakistan.

The following (see Fig. 2) shows the existing and required power-interest map. Stakeholder 1 government sector with high power still has low interest in managing plastic waste or reclaiming its energy or material value, but as we can see from many

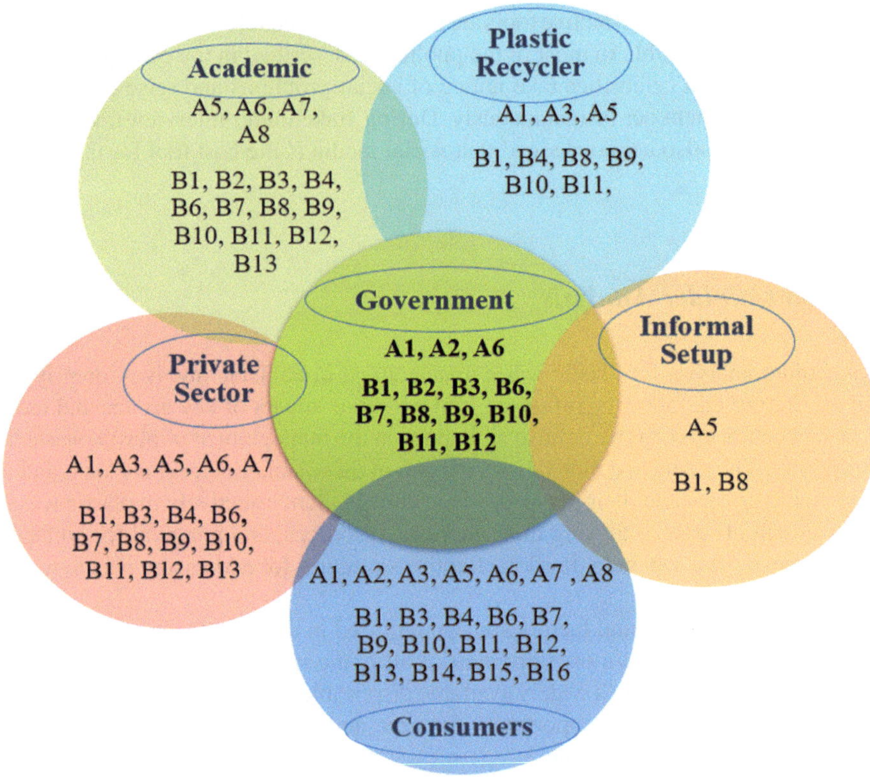

Fig. 1 Stakeholder's role in public awareness tools

public awareness tools for education and action, the government's involvement and representation is required for positive outcomes. 2-The academia sector even with low power, is doing a better job but can be improved if the government sector puts more investment into capacity building. 3-Plastic Producers similarly are high in power as they contribute to the problem and can afford to take their corporate social responsibility and pay for pollution.

Stakeholder 4-Plastic Recyclers, and 6-Consumers both with low power are actively interested in the issue. But there is nothing much they can do without the government's adequate support. Ironically, 5-Informal Sector contributing the most to solving Pakistan's plastic disaster is less in power and less interested in raising awareness about the issue. In fact, as per interviewees' comments, people working in this sector are reluctant to disclose their status. It is one of the reasons that none of the fifteen shortlisted interviews conducted for this study belong to this sector. The government bodies remain at the centre of all activities as it is highly encouraged that government support is still missing, as identified from the recycler interviewer's comments.

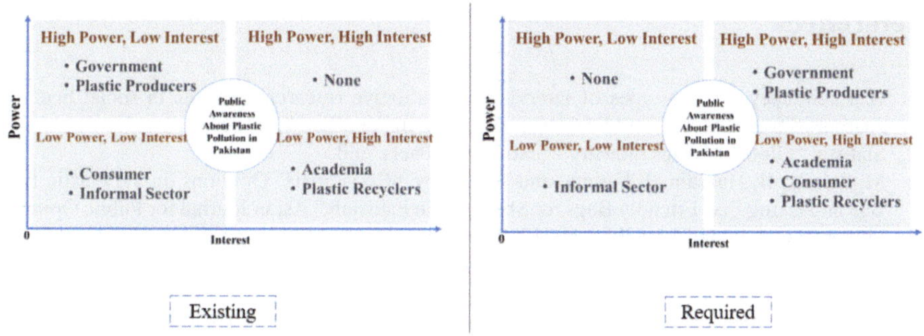

Fig. 2 Existing and required power-interest map

4 Conclusion

The country is struggling with plastic pollution, with the River Indus ranking as the second most plastic-polluted river in the world. Bans on disposable plastics have been ineffective, and the lack of awareness and operational recycling facilities have exacerbated the problem. The study aims to identify public awareness tools, interview relevant stakeholders, and develop a stakeholder engagement analysis to achieve effective plastic waste management in Pakistan.

Throughout the study, the focus was on gathering primary data from experts in the PWM sector. The study applied a semi-structured interview type with predetermined open-ended questions to facilitate the conversation between interviewees and interviewers. Additionally, a token of appreciation was given to the interviewees in the form of a donation to a cancer hospital in the name of the interviewees.

In such a scenario, these temporary alternatives are only increasing the complexity. PET is the only plastic type which is widely being recycled in the country but that is attributed only to the efforts of informal sector and junk dealers. However the public needs constant reminders and awareness to motivate them to act. This study has identified six key stakeholders; 1-Government, 2-Academia, 3-Plastic Producers, 4-Plastic Recyclers, 5-Informal Sector, and 6-Consumers and analysed their power vs interest on a power-interest map. The government, despite its high power, shows low interest in addressing plastic waste, while other stakeholders like academia and plastic producers are more engaged, yet need greater support from the government for capacity building and corporate responsibility; similarly, low-power groups such as recyclers and consumers are interested but lack the means to act without government intervention, highlighting the crucial role of government in achieving positive outcomes.

References

1. H. L. Stuckey, "Three types of interviews: Qualitative research methods in social health," Journal of Social Health and Diabetes, vol. 1, no. 02, pp. 56–59, 2013.
2. Statista, "Global plastics industry - Statistics & Facts."n.d.
3. M. Zaheer, B. Hussain, T. Fatima, and A. Edgley, "Consumers' Opinions on the Plastic Bag Ban and Using Eco-Friendly Bags for Shopping in Pakistan," Asian Journal for Public Opinion Research, vol. 9, no. 2, pp. 161–187, 2021.
4. I. Nishimura, "Strategy for Plastics in a Circular Economy," Seikei-Kakou, vol. 30, no. 11, pp. 577–580, 2018.
5. Malik Amin Aslam Khan, "To eliminate plastic pollution, banning plastic bags is not enough—creative solutions are needed." Accessed: Oct. 11, 2024. [Online]. Available: https://www.weforum.org/agenda/2021/11/4-ways-pakistan-is-tackling-plastic-waste-and-pollution/
6. Z. Gülenç, D. Konaklı, and İ. D. Öztürk, "Raising awareness about marine litter through beach cleanup activities along the Turkish coasts of the Black Sea," Marine Litter in, p. 74, 2020.
7. "Coastal cleanup drive conducted to conserve marine life, reduce plastic pollution," The News, Karachi, Apr. 28, 2018.
8. L. M. Heidbreder, J. Steinhorst, and M. Schmitt, "Plastic-Free July: An Experimental Study of Limiting and Promoting Factors in Encouraging a Reduction of Single-Use Plastic Consumption," Sustainability 2020, Vol. 12, Page 4698, vol. 12, no. 11, p. 4698, Jun. 2020, doi:https://doi.org/10.3390/SU12114698.
9. N. Yuliastuti, E. T. Sulistyo, and N. Yulianto, "Plastic as A Media of Installation Artworks Based on Recycle Art," pp. 65–71, Feb. 2019, doi:https://doi.org/10.2991/ICALC-18.2019.10.
10. F. S.H. and A. P., "CHALLENGES IN 3R IMPLEMENTATION AND PUBLIC PARTICIPATION," Malaysian Journal of Science, vol. 32, no. 2, pp. 49–58, Dec. 2013, doi:https://doi.org/10.22452/MJS.VOL32NO2.5.
11. J. J. Klemeš, Y. Van Fan, and P. Jiang, "Plastics: friends or foes? The circularity and plastic waste footprint," Energy Sources, Part A: Recovery, Utilization, and Environmental Effects, vol. 43, no. 13, pp. 1549–1565, Jul. 2021, doi:https://doi.org/10.1080/15567036.2020.1801906.
12. M. Jahanzaib, G. Fatima, and D. e Nayab, "Review of Single National Curriculum with Perspective of the Education of Children with Visual Impairment at Primary Level in Punjab Pakistan," Journal of Business and Social Review in Emerging Economies, vol. 7, no. 3, pp. 547–560, 2021.
13. T. Stanton et al., "Planet Patrolling: A citizen science brand audit of anthropogenic litter in the context of national legislation and international policy," J Hazard Mater, vol. 436, p. 129118, Aug. 2022, doi:https://doi.org/10.1016/J.JHAZMAT.2022.129118.
14. E. M. Bennett and P. Alexandridis, "Informing the Public and Educating Students on Plastic Recycling," Recycling 2021, Vol. 6, Page 69, vol. 6, no. 4, p. 69, Oct. 2021, doi:https://doi.org/10.3390/RECYCLING6040069.
15. Lala Rukh Memon, "Let's Fix the Marine (Non-Recyclable) Plastic Litter Issue!," Lindau Nobel Laureate Meetings.
16. Unilever, "The year 2050 will see more plastic in the ocean than fish.," Dawn, Karachi, Feb. 2021.
17. Correspondent, "Celebrities join hands to #BeatPlasticPollution," Express Tribune, karachi, Aug. 18, 2018.

Applying a Transdisciplinary Approach to the Development of a Biologically Circular Beanie

Line Fournier, Larissa Hänni, and Maria A. Franco

Abstract The problem of textile waste is enormous. In Switzerland, about 11 kg of clothes are discarded per person annually. Compostable textiles may offer a solution to the significant amount of waste nowadays. However, there are several barriers to developing and disposing of biologically circular textiles, including the fact that the industrial composting of textiles is not yet established. The present paper aims to comprehensively address the development of a biologically circular beanie by integrating a transdisciplinary approach to tackle ecosystem considerations. In order to answer the research questions, a stakeholder analysis, as well as a quantitative and qualitative analysis were carried out. Final results indicate that the project presents both risks and opportunities, particularly in relation to the system, supply chain, production capacity, and customer acceptance. The study implies that regulatory changes are necessary in Switzerland to enable the composting of textiles. Additionally, a product passport could aid in determining the compostability of the beanie. It is also recommended to engage in transdisciplinary collaboration to find a practical solution that satisfies all stakeholders.

Keywords Circular economy · Circular design · Transdisciplinary

---

L. Fournier (✉) · L. Hänni
Bern University of Applied Sciences, Bern, Switzerland
e-mail: line.f@hotmail.com

L. Hänni
e-mail: larissa.fabienne.haenni@alumni.bfh.ch

M. A. Franco
Bern University of Applied Sciences, Biel, Switzerland

A. Podleisek et al. (eds.), *Advances in Circular Economy: Bridging Research and Practice*, Springer Proceedings in Earth and Environmental Sciences,
https://doi.org/10.1007/978-3-032-21547-5_4

1 Introduction

Triggered by low-cost fast fashion and an enlarged middle-class worldwide, the textile industry is currently facing a major waste problem. Worldwide, about 92 million tonnes of textile waste are produced every year [1]. In Switzerland, the situation is no different. Each year, the average Swiss citizen buys 16 kg of new textiles per year and throws away 11 kg in recycling collection bins located all around the country [2]. Up to two-thirds of the textiles collected in Switzerland, however, are still in good enough condition to be worn. Collected clothing is mostly sold abroad, mainly in Eastern Europe, Russia, Africa, and the Middle East [3]. Kenya is as one of the main recipient countries of Swiss textile waste [4] and one of the largest importers of used clothing in Africa, collecting about 185,000 tons of used clothing each year. Only about 10 percent of discarded clothing is collected for reuse in Switzerland [2].

The textile industry faces a growing environmental crisis, with the majority of textile waste composed of non-biodegradable synthetic fibres like polyester and polyamide/nylon [4]. These synthetic materials, derived from non-renewable resources, not only require energy-intensive production processes but also leave a lasting environmental footprint throughout their lifecycle. The environmental impact of synthetic fibres is far-reaching and multifaceted. From clogging landfills to polluting oceans with microplastics, these materials pose significant threats to ecosystems worldwide [5]. In response to these challenges, the textile industry is witnessing a shift towards more sustainable alternatives. Natural fibres are gaining popularity as manufacturers seek to mitigate the environmental damage caused by synthetics [6]. Wool, for example, stands out as a fully biodegradable option in both terrestrial and marine environments. Depending on soil conditions, climate, and specific wool characteristics, it can biodegrade within a remarkably short span of three to four months [7].

In this paper we differentiate between two terms: biodegradability and compostability. Biodegradation is a natural degradation process that occurs when microorganisms break down organic substances into simpler molecules [8]. This process can take place in various environments such as soil, water, landfills, and compost heaps. The rate and extent of biodegradation depend on factors like temperature, humidity, and the presence of specific microorganisms. On the other hand, a compostable product is made from biodegradable material that decomposes within a specified time, in a controlled environment, and without toxicity to the environment [9]. This process occurs in composting facilities or home composting systems, resulting in the production of compost or humus, a stable organic material beneficial for soil enrichment [10].

This study is aimed at exploring the potential benefits of applying a transdisciplinary approach to the development of circular products, particularly focusing on the creation of a biologically circular beanie. Transdisciplinarity in the context of sustainability integrates knowledge from various academic disciplines while also incorporating insights from non-academic stakeholders, such as community members, policymakers, and practitioners, emphasizing the importance of collective understanding and the co-creation of knowledge [11]. We posit a transdisciplinary approach, one that includes multiple scientific disciplines focusing on shared problems and the active input of practitioners from outside academia [12]. The aim is to contribute to the research by evaluating a practical case study and to give perspectives for a new EOL scenario that could partially mitigate the problems of textile waste. It also seeks to explore the concrete implications that implementing the biological cycle in the textile industry could have, considering the views of the main stakeholders for such a product.

One key question guiding this study is: "How can the application of a transdisciplinary approach help in the development of a circular product?" Additionally, the study seeks to examine the risks and opportunities associated with the development of such a product. Thus, another crucial question driving this research is: "What are the risks and opportunities when developing a biologically circular beanie, considering systemic readiness, customer acceptance, and supply chain capability?" By addressing these research questions, this study aims to provide valuable insights into the feasibility and implications of transitioning towards a more circular and sustainable textile economy.

This research was conducted in cooperation with a practical partner working in the fashion industry. The Swiss company is specialized in the design and sale of headwear products, with a strong focus on winter and outdoor sports. The products are designed in-house, but the production is outsourced to their longstanding business partner in Poland. Most of their products are sold directly in their online shop (B2C), but they also sell their products in partner stores in Switzerland and in other shops in Europe or Japan.

2 Methods

To address the proposed research questions, we employed a multi-method approach from October 2023 to January 2024. Our methodology consisted of the following components: (i) stakeholder analysis, (ii) survey distribution, (iii) interviews, and (iv) focus groups. Finally, the results were summarized in a risk map. Figure 1 provides an overview of the applied methods in the project, timed by standardized calendar weeks, whereas the upcoming subsections provide a brief overview of the employed multi-method approach.

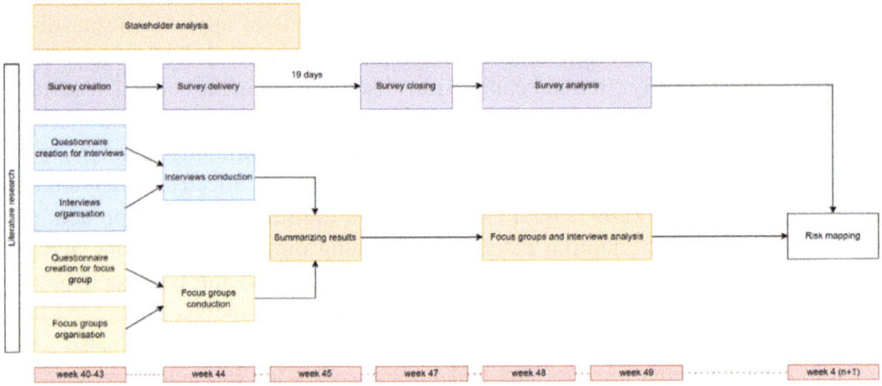

Fig. 1 Overview over applied method in the project

2.1 Stakeholder Analysis

A stakeholder analysis was carried out to identify the key actors in the field and to select participants for interviews and focus groups. This analysis aimed to identify individuals, groups, and organizations affected by decisions or actions related to our research [13]. The process unfolded in three stages. In the first stage, we engaged with our partner company to gain preliminary insights and obtained initial contacts from them. In a second stage, we expanded our search by utilizing Google to find additional relevant contacts, focusing on answering the following questions: (i) what companies or experts have knowledge of biodegradable or compostable textiles in Switzerland?, (ii) which stakeholders are currently involved in the collection, sorting, and composting process for textiles? and (iii) which suppliers can provide suitable raw materials for the development of a biologically circular beanie? In a third stage of the stakeholder analysis, we leveraged ongoing interview data to further refine and expand our list of contacts. After compiling a comprehensive inventory of stakeholders, we employed the power/interest matrix developed by Reed et al. [13] to categorize and prioritize them. This matrix allowed us to visualize and analyse the relative influence and interest levels of each stakeholder, providing valuable insights into their potential roles and importance in our research.

2.2 Survey

A survey was conducted to gather insights from a wide range of current and potential customers about their habits with beanies and their acceptance in terms of biodegradable/compostable material and price. The aim was to collect responses from a wide range of consumers, with a focus on people aged between 25 and 35 living in Switzerland. This age group was chosen because it is the partner company's target market and is most interested in beanies. The online survey was developed using LimeSurvey and made available in the three main languages relevant to the target groups: French, German, and English. By offering the survey in Switzerland's two official languages, French and German, as well as English, we ensured comprehensive coverage and accessibility for the entire target audience. The translations were cross checked by members of the research team who were native in each of these languages. The questionnaire consisted of 40 questions, including closed and multiple-choice questions. The questionnaire was divided into the following sections: (i) general, (ii) clothing purchasing habits, (iii) recycling and composting, and (iv) the use and purchase of a beanie. The average time to complete the questionnaire was from 7 to 10 min. The survey was reviewed within the research group and validated by the partner company before launching the final survey. The survey was disseminated in the authors' social networks (i.e., WhatsApp, Instagram, and LinkedIn), in the partner company's newsletter, and by email to all BFH-HAFL (School of Agriculture, Forest, and Food Science) students and employees. The online survey was accessible for a period of 19 days, from November 1 to November 19, 2023. We received a total of 500 valid responses during this timeframe. After carefully filtering the participants to ensure they met our target group criteria, we refined the final sample to 490 responses. To analyse the dataset, we employed a combination of analytical tools, including Excel, LimeSurvey, and RStudio. Our analysis primarily focused on descriptive statistics, allowing us to extract meaningful insights from the survey responses and paint a comprehensive picture of our research findings.

2.3 Focus Groups

To facilitate an interactive and transdisciplinary discussion among stakeholders, we organize two focus groups. This method was chosen to explore diverse, in-depth opinions and stimulate discussions between different parties, encouraging the cross-pollination of ideas. The focus groups aimed at investigating the current market situation of textiles, assessing knowledge about biodegradable/compostable materials, exploring current end-of-life scenarios for textiles, and identifying areas requiring further research. The first focus group, held on November 1st, 2023, consisted of three participants: an expert in sustainable development, a supplier of sustainable textiles, and an expert from the composting industry specializing in end-of-life scenarios. The second focus group, conducted on November 3rd, 2023, included two textile experts with both practical and academic backgrounds. Participants were carefully selected based on the previously conducted stakeholder analysis. Both focus groups were held online in English via Microsoft Teams and were recorded for subsequent analysis. The research team assigned roles of moderator and note-taker for each session. After introducing the participants and the research team, we presented the focus group topics and posed semi-structured questions to encourage discussion. The first focus group lasted 1 h and 10 min, while the second lasted 50 min.

2.4 Interviews

To gain a more comprehensive understanding of a research topic, we conducted a series of five semi-structured interviews with industry experts. We chose the semi-structured interview format to allow for flexibility and spontaneity during conversations, enabling us to explore emerging themes and insights. The experts were selected based on their professional expertise in key areas of our study, including circular textiles, textiles repair processes, and the production, design, and disposal of biodegradable/compostable textiles. We identified these contacts through recommendations from recommendations from our practical partner as well as from focus group participants, and targeted industry research on Google. The selection process was guided by our stakeholder analysis to ensure a diverse and relevant pool of interviewees. Each interview was conducted online via Microsoft Teams and lasted between 30 and 50 min. To complement these expert interviews, we also held regular meetings with our practical partner, which provided additional insights and information, further enriching our understanding of the field and ensuring our research remained aligned with industry perspectives and needs.

2.5 Written Communication

Individuals who were unable to attend the focus groups or interviews were contacted via email to answer questions. Informants from all the qualitative methods have been listed and numbered, and can be found in the Appendix.

To ensure the privacy of the individuals that attended our study, we anonymized the data, using the abbreviation "INF" followed by a number.

3 Results

3.1 Application of a Transdisciplinary Approach to the Development of a Biologically Circular Beanie

The stakeholder analysis identified 12 main categories of stakeholders important to the development of a biologically circular beanie and more specifically through the following selected life-cycle stages: design, manufacturing, distribution & sales, use, end-of-life. Depending on their role in the project, their power and interest differ and position them accordingly on the matrix in Figure 2. Depending on their level of involvement as defined by their level of power and interest, the approach to be adopted towards these stakeholders varies from a minimal effort of monitoring for secondary stakeholders to more substantial interventions such as the engagement of key stakeholders.

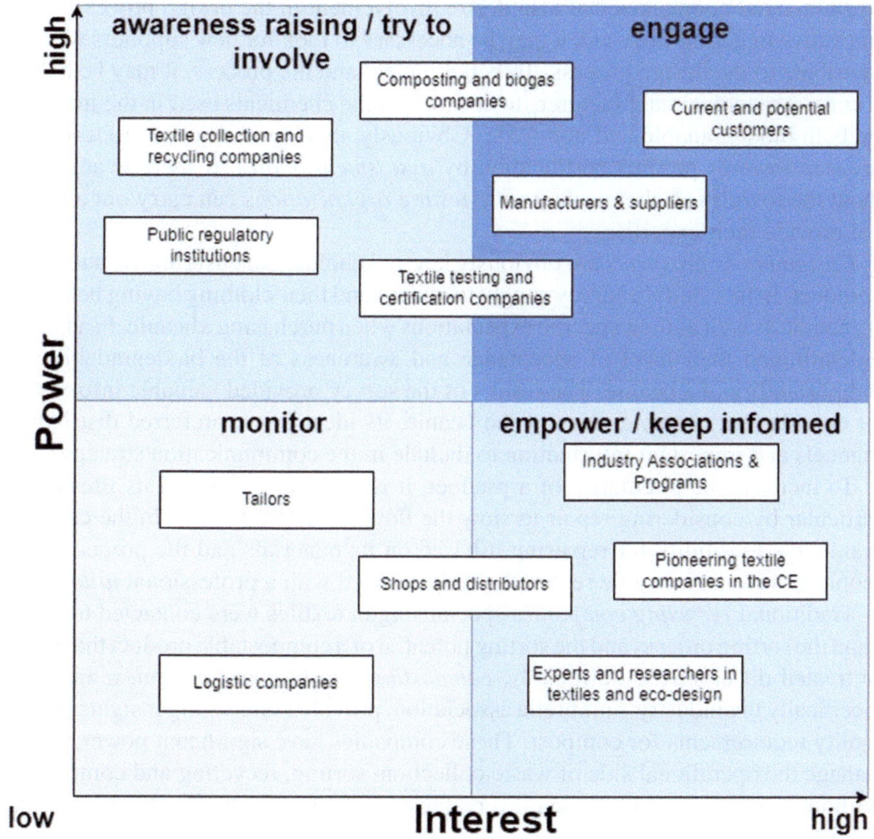

Fig. 2 Power/interest matrix (based on [13]).

The various methods cited in Chap. 2 have highlighted the practical implications for each stakeholder in the development of such a product, as well as identified the main risks and opportunities.

First, in the exploration and development phase of an atypical product, *experts and researchers* specialising in textiles or eco-design provided important insights to guide the initial research. *Pioneering companies* in the field of circular and sustainable textiles, as well as *industry associations and programmes*, set standards, hold key knowledge and experience, and inspire and promote circularity. Knowing their points of view means not starting research completely from scratch and benefiting from their expertise. Although these stakeholders have no decision-making power in the development of the beanie, it is important to keep them informed so that research, and the industry as a whole, can evolve.

Manufacturers and suppliers then provide the materials and assemble the beanie. In the case study, the practical partner heavily relied on these main manufacturers, with whom it had been working for several years. Generally, they are small to medium-sized companies that also desire involvement in the design process. When innovative materials are used, it may be necessary to look for new suppliers who can contribute to the design process. To better understand the process, it may be helpful to contact machine manufacturers to ensure that the chemicals used in the machines are both biodegradable and non-toxic. Obviously, this knowledge of processes is a key step towards product certification by *assessment bodies*. If there is any doubt about the toxicity of chemicals, *textile testing organisations* can carry out analyses and provide their expertise.

Customers or prospects are obviously key stakeholders in ensuring the success of a product. In this study, a survey sought to understand their clothing buying behaviour in general, as well as their specific expectations when purchasing a beanie. In addition, it determined their level of acceptance and awareness of the biodegradability or compostability of a beanie. The results of the survey provided valuable information for defining the characteristics of the beanie, its ideal price, preferred distribution channels and important information to include in the communication strategy.

To increase the circularity of a product, it is essential to extend its lifespan, in particular by considering repair to slow the flow of materials [14]. In the case of a beanie, the feasibility for repairing it based on its materials and the propensity for people to have their beanie repaired were discussed with a professional *tailor*.

Traditional *recycling companies* specialising in textiles were contacted to understand the sorting process and the sorting potential of a compostable product that would be treated differently. Additionally, *composting and biogas companies*, and more specifically the industry's umbrella association, provided interesting insights into the quality requirements for compost. These companies have significant power, as they manage the operational side of waste collection, sorting, recycling and composting. Without a consensus with these players, the end-of-life scenario for the beanie may differ from the initial plan.

If collection and/or industrial composting cannot be guaranteed, the company should consider taking back the end-of-life products itself to ensure separate composting on a smaller scale. Consequently, this requires a *logistics company* to guarantee the return of the product. In this case, *regional public institutions* should allow a derogation from the Ordinance on the Limitation and Disposal of Waste, which normally requires municipalities to collect waste through specialised textile waste collection companies.

Finally, *shops and distributors* can be consulted about their interest in selling and promoting such a product.

The different categories of stakeholder play different roles throughout the beanie's life cycle and have a different importance during its development (see Figure 3). Referring to the stakeholders positioned in Fig. 2, they are then classified in Fig. 3 according to the stage at which they intervene in the beanie's life cycle. Interestingly, stakeholders with a higher level of interest or power tend to be involved in the early stages of the life cycle. However, the intermediate and final stages of the life cycle are no less important and may even be decisive in guaranteeing the biodegradability of the beanie. The power of certain players, such as composting and biogas companies,

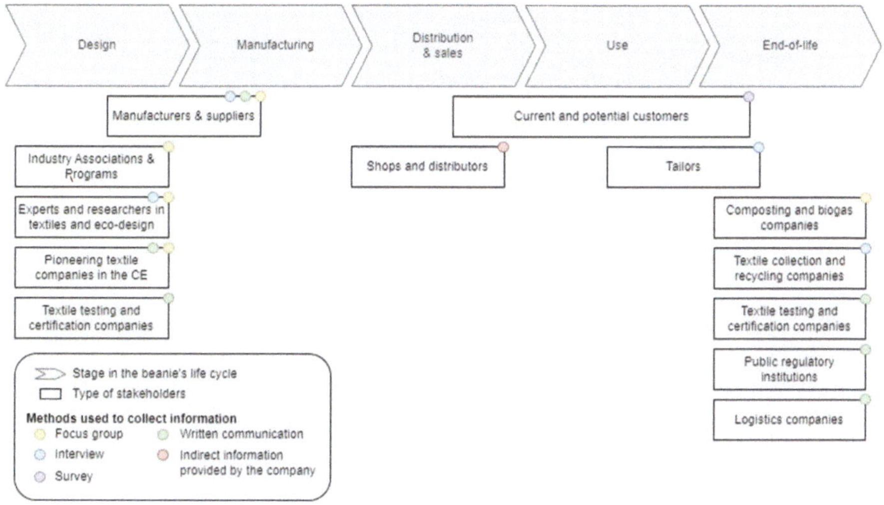

Fig. 3 Stakeholder engagement and information gathering across product life cycle stages

is reflected in Figure 2, and may highlight points for attention at particular stages of the life cycle (Fig. 3).

In an interdisciplinary approach, it is important to collect the points of view of each of the stakeholders identified earlier to obtain an overall view of the opportunities for developing such a product or, conversely, of the barriers and risks that could prevent it from progressing smoothly. Similarly, it is essential to consider the realities of stakeholders, from the design phase and throughout the life cycle, to ensure the feasibility of a circular scenario, such as composting or biodegradation of the product.

3.2 Knowledge Production in Relation to Risks and Opportunities

Through the different methods used, the risks and opportunities across the entire product life cycle and for the relevant stakeholders were identified. Figure 4 compiles and interrelates these risks and opportunities. Some aspects are broader and focus on textiles and clothing in general, while others are more specific and relate to the direct development of the biologically circular beanie.

Three main risks and opportunities have been identified, i.e. those related to the system, those related to supply chain and production capacity and finally, those related to customers.

(i) System in Switzerland—systemic readiness

The focus groups confirmed that there are currently no industrial composting facilities for textiles in Switzerland. Additionally, it is not possible to certify a garment

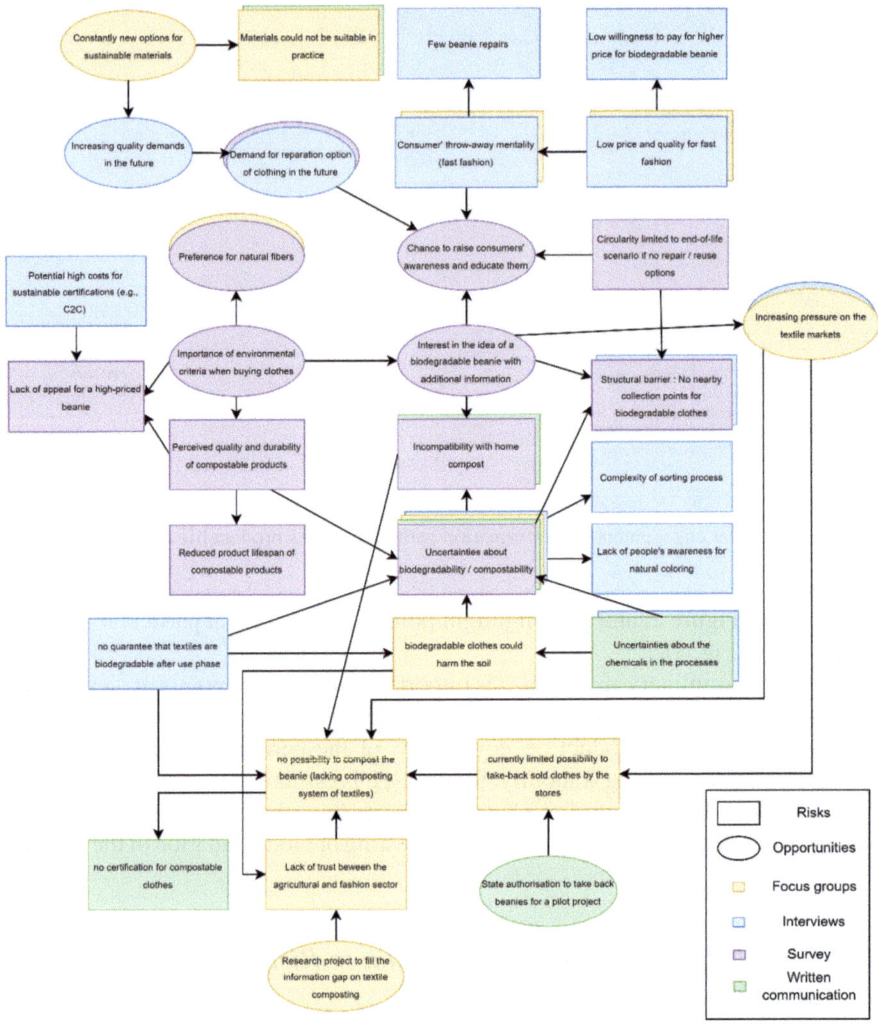

Fig. 4 Risk and opportunity mapping

as compostable since this is not the end-of-life scenario for which it is typically intended (INF22). One alternative would therefore be home composting, but in the survey several respondents raised concerns, particularly about the quality of the final compost and the feasibility of making it decompose with ordinary green waste. If the company decides to set up its own composting programme on a smaller scale, it would have to offer a take-back programme for the beanies sold. This would involve a derogation from Article 13 of the Ordinance on the Limitation and Disposal of Waste. The competent regional public institution seemed open to considering this as part of a pilot project (INF19). Returning the product would also entail an additional cost to be paid to a logistics company (INF18) and could have an impact on customer acceptance if the cost is included in the price.

For recyclers, one of the biggest challenges is sorting textiles, whose composition is very heterogeneous (INF14) and makes the process complex. One concern is being able to easily ascertain the biodegradability or compostability of the product and the chemicals it contains, while also considering that some customers may have used environmentally toxic detergents that have impregnated the fibre during the use phase (INF06). In addition, there are uncertainties linked to the decomposition time of textile waste compared to ordinary green waste. If the decomposition time is different, it is possible that the textiles could hinder the process and get stuck in the biogas machine, for example (INF06). Recyclers and experts have therefore suggested creating a separate route for processing biodegradable or compostable textile waste (INF06 & INF08), or the exploration of biochar as a safer solution for minimising the risks associated with chemical products.

One industry association has expressed an interest in composting textiles and is undertaking research with several players to prove the feasibility as well as the environmental benefits that can be derived, particularly in terms of the valuable nutrients for the soil available in certain materials (INF04). However, there is a lack of scientific research in this area. One of the biggest barriers is also the lack of trust between players in different industries, particularly between the agricultural sector, which uses compost, and the textile sector. INF03, the umbrella association for composting companies, confirmed these fears about the impact on the soil: "*We do not allow anything that could have a negative effect on the soil. So if there is a risk of having microfibres accumulating in the soil and all that stuff, then it wouldn't be allowed, although it is certified as biodegradable. These are not really industry standards as such, there is rather a branch standard for agriculture and the waste branch. In Switzerland a certain level of quality has to be in guaranteed in material that is brought into or ends up in the soil. It is all about soil quality and fertility*".

Substantial changes need to be made to the system to ensure the success of this end-of-life scenario. Although it is still uncertain whether such a composting route for textiles will ever come to fruition (INF08), INF09 stressed the importance of initiating this change even if the system does not yet exist. INF02 also emphasised the need to take greater account of the biological cycle and draw inspiration from nature.

In general, opportunities for research and collaboration between industries seemed to be at hand and necessary to push the system forward. This appeared all the more relevant given the pressure on the textile market to adopt more sustainable practices and the arrival of new regulations.

(ii) Production process—supply chain capability

More and more materials are emerging for biodegradable or compostable textiles, offering new opportunities for product development. For example, natural or artificial fibres that are intrinsically biodegradable have been recommended, along with more innovative materials such as bio-based and biodegradable polymers. The focus group I concluded that no single solution is recommended. Thus, even if cotton is naturally biodegradable, it cannot be the only material used in clothing (INF01). When using atypical materials, manufacturers must also ensure that the materials are actually compatible with each other and with the machines (INF24).

Throughout the life cycle, the use of chemical products is also a critical aspect to consider, in particular to guarantee non-toxicity to soil. However, contacts with suppliers and manufacturers have revealed a lack of knowledge about the chemicals used, leading to many uncertainties. When chemicals are used in machines, the risk of them impregnating the fibre is not known or quantified. More research needs to be carried out throughout the supply chain to ensure that chemicals are used that have no impact on biodegradability, requiring time, transparency and trust between the players. This also applies to any repairs, which should only use materials that are also biodegradable or compostable.

(iii) Customers—customer acceptance

The survey revealed that the ecological footprint of a garment was very or extremely important for 51.22% of participants. Overall, the importance of ecological criteria obtained a score of 3.5/5 (1: not at all important, 5: very important), i.e. moderately to very important. This reflects the environmental awareness of a majority of participants. Among the most important criteria for considering a beanie to be sustainable, participants mentioned the use of organic or recycled materials, or sustainable and ethical manufacturing processes. For headwear products, 61.63% of participants preferred natural fibres. However, when it came to buying a sustainable beanie, biodegradability was considered to be the least important criterion. Surprisingly, over ¾ of participants would still be interested in buying a compostable beanie. However, this interest seemed to be conditional for some and could be increased by the provision of additional information. Indeed, participants expressed an interest in obtaining information on the composting process, the materials used, care and composting instructions and finally, the real environmental benefits. The most reluctant participants also expressed doubts and uncertainties about the possibilities of composting at home and the impact on the quality of the final compost, as well as the durability of the beanie and the risk of premature decomposition, or alteration of the technical characteristics (e.g. water-resistance). These insights provide an opportunity to educate interested customers and raise their awareness. Obviously, to increase customer acceptance, it is essential to include these elements transparently in the company's communication strategy.

According to INF07, one aspect to consider is the throw-away mentality that has increased with fast fashion. Although, in principle, anything can be repaired regardless of the material, repairs to knitwear are typically visible, and it is uncommon for people to have beanies repaired (INF07). However, a nearby, free or affordable service could encourage participants to have their clothes repaired. According to their responses, 92.45% of them repaired their clothes, but only 20% did so systematically. Despite this, 81.43% of participants seemed to keep their beanies for at least 5 years. The fairly long lifespan of this product led one participant to question the relevance of making such an item compostable, in line with a comment by INF03, which would advocate this feature for rather short-lived products.

For 52.65% of respondents to the survey, price was one of the top 3 most important criteria when buying clothing. For a sustainable beanie, 74.90% of respondents would be prepared to pay 3 to 10 Swiss francs more than for a standard beanie, for which a third of respondents would pay between 26 and 35 Swiss francs. In assessing the feasibility of industrial composting as a prospective option, it was determined that the vast majority of participants would be willing to go to a centre to dispose of old clothes, but often on condition that the centre is close to their home. Nevertheless, 52.45% of respondents indicated that they did not know whether a composting centre existed in their area.

4 Discussion

In conclusion, when developing a circular product, the use of a transdisciplinary approach appears to provide valuable insights into maximising circularity. This approach includes an analysis of stakeholders throughout the product life cycle, as recommended by the principles of eco-design. This transdisciplinary approach identified a number of risks and opportunities in the main categories of system readiness, supply chain capacity, and customer acceptance. For instance, composting companies have expressed reluctance to include textile waste in compost because of its toxicity for the soil, while recycling companies are concerned about the complexity of the sorting process. Producers also expressed doubts about the chemicals that could affect the biodegradability and compostability of the final product, and consumers requested additional information before confirming their interest in purchasing such a product. The results of the transdisciplinary approach concluded that the development of a compostable beanie was not advisable at this stage but identified opportunities for collaboration in the development of a future product, and for advancing research and knowledge in this field. The information to be provided to reduce customer doubts has also been identified and can be used not only for future product development but also for educational purposes.

Overall, there is still considerable room for progress and development in implementing of such an end-of-life scenario for textiles. However, certain barriers could be addressed through regulations or facilitators to encourage further advancement. For instance, the introduction of a digital passport for products could simplify clothing sorting processes, while political support might be pivotal in launching small-scale pilot projects. Businesses and industries can also play a crucial role in reducing the impact of textiles by fostering cross-sector collaborations. These collaborations can help test the technical and economic feasibility of composting textiles and demonstrate their potential environmental benefits.

Future research avenues, both practical and academic, hold significant promise. Other studies could explore the perspectives of various stakeholders across different regions and scales. This is particularly important as this study focuses on Switzerland and contains a clear bias due to its association with a single company. Indeed, certain informants were chosen due to their affiliation with the company, and the survey sample includes its customers.

Appendix

Abbreviation	Category	Type of qualitative method
INF00	Company	Interview/written communication
INF01	Expert in textiles	Focus group
INF02	Pioneering textile company in the CE	Focus group
INF03	Composting and biogas company	Focus group
INF04	Industry association & program	Focus group
INF05	Researchers in textiles	Focus group
INF06	Collection and recycling company	Interview
INF07	Tailor	Interview
INF08	Expert in textiles	Interview
INF09	Expert in eco-design	Interview
INF10	Supplier/researcher	Interview
INF11	Supplier/manufacturer	Written communication
INF12	Supplier	Written communication
INF13	Supplier	Interview
INF14	Textile collection and recycling company	Interview
INF15	Supplier	Written communication
INF16	Supplier	Written communication
INF17	Supplier	Written communication
INF18	Logistic company	Written communication
INF19	Regional public institution	Written communication
INF20	Machines manufacturer	Interview
INF21	Recycling company (start-up)	Written communication
INF22	Certification company	Written communication
INF23	Supplier	Written communication
INF24	Manufacturer	Interview/written communication

References

1. Harris, L. (2023, August 17). Here's how textile recycling can create jobs and reduce pollution. World Economic Forum. https://www.weforum.org/agenda/2023/08/textile-recycling-create-jobs-reduce-pollution/
2. Circle Economy. (2023). Circularity Gap Report Switzerland. https://assets-global.website-files.com/5e185aa4d27bcf348400ed82/647f3580e21fe81c006837ea_20230525%20-%20CGR%20Switzerland%20-%20Report%20-%20210x297mm.pdf. p. 43.
3. Swiss Federal Office for the Environment. (2019, July 5). Kleider und Schuhe. Swiss Federal Office for the Environment. https://www.bafu.admin.ch/bafu/de/home/themen/thema-abfall/abfallwegweiser%2D%2Dstichworte-a%2D%2Dz/kleider-und-schuhe.html
4. Sandmeier, M. (2022, May 5). Fast fashion: De fausses promesses de durabilité. Greenpeace Suisse. https://www.greenpeace.ch/fr/story-fr/85027/fast-fashion-fausses-promesses/.
5. European Environment Agency. (2021, January 28). Plastic in textiles: Towards a circular economy for synthetic textiles in Europe — European Environment Agency [Briefing]. European Environment Agency. https://www.eea.europa.eu/themes/waste/resource-efficiency/plastic-in-textiles-towards-a
6. Chen, X., Memon, H. A., Wang, Y., Marriam, I., & Tebyetekerwa, M. (2021). Circular Economy and Sustainability of the Clothing and Textile Industry. Materials Circular Economy, 3(1), 12. https://doi.org/10.1007/s42824-021-00026-2. p. 12.
7. The Woolmark Company. (n.d.). Biodegradable fashion: Breathe easy with wool. Woolmark. Retrieved 12 January 2024, from https://www.woolmark.com/environment/biodegradable-fashion-breathe-easy-with-wool/.
8. Hodzic, A. (2004). 12—Re-use, recycling and degradation of composites. In C. Baillie (Ed.), Green Composites (pp. 252–271). Woodhead Publishing. https://doi.org/10.1016/B978-1-85573-739-6.50015-4.
9. TÜV Austria Belgium. (n.d.). Biodegradable vs compostable. https://www.tuv-at.be/fileadmin/user_upload/docs/download-documents/SK/PD-BA-TABE-CERT-BIO-SK-01_biodegradable_vs_compostable_EN.pdf
10. Rudnik, E. (2008). Compostable polymer materials (1st ed). Elsevier.
11. Kudo, S., Omi, K., Florentin, K., & Allasiw, D. I. (2021). Key experiences for the transdisciplinary approach: Fieldwork-based training in sustainability science education. International Journal of Sustainability in Higher Education, 22(3), 615–634. https://doi.org/10.1108/IJSHE-05-2020-0185.
12. Brandt, P., Ernst, A., Gralla, F., Luederitz, C., Lang, D. J., Newig, J., Reinert, F., Abson, D. J., & von Wehrden, H. (2013). A review of transdisciplinary research in sustainability science. Ecological Economics, 92, 1–15. https://doi.org/10.1016/j.ecolecon.2013.04.008. p. 1.
13. Reed, M. S., Graves, A., Dandy, N., Posthumus, H., Hubacek, K., Morris, J., Prell, C., Quinn, C. H., & Stringer, L. C. (2009). Who's in and why? A typology of stakeholder analysis methods for natural resource management. Journal of Environmental Management, 90(5), 1933–1949. https://doi.org/10.1016/j.jenvman.2009.01.001.
14. Konietzko, J., Bocken, N., & Hultink, E. J. (2020). Circular ecosystem innovation: An initial set of principles. Journal of Cleaner Production, 253, 119942. https://doi.org/10.1016/j.jclepro.2019.119942.

A Holistic Framework Proposal for Circular Construction from Current Trends, System Needs, and Sustainable Objectives

Roger Vergés, Kàtia Gaspar, David Font, and Núria Forcada

Abstract The adverse effects of the built environment on ecosystems are aggravated by the looming threat of climate change impacts on society. Given that the construction sector significantly contributes to emissions, a shift is imperative to curtail its expansion and mitigate environmental and social impacts. This paper aims to analyse the current trends, system needs, and sustainable objectives within the construction sector from a location-specific perspective to identify flaws, unaddressed issues, bottle necks, and areas for improvement. To that end, sustainability and circularity metrics, trends in decision-making processes, the utilisation of Building Information Modelling (BIM), its integration with complementary technologies, and a state-of-the-art holistic assessment of buildings throughout their entire life cycle are examined. Once evaluated, a comprehensive framework proposal is outlined to address these topics, with a priority on promoting circular economy (CE) practices, particularly circular degrowth, with the aim of alleviating the adverse effects of the construction sector on the climate. Findings show that (i) circularity metrics are seldom incorporated into LCSA processes, provoking lack of information, (ii) automated decision-making criteria are infrequently utilised in construction projects, (iii) data visualisation remains uncommon in BIM environments, and (iv) product selection should not be conducted in isolation but rather should consider entire system typologies within the project. These findings, along with the proposed holistic framework, present an opportunity to transform the built environment into a more sustainable entity.

Keywords Circular economy · Circular construction · Sustainability · Building information modelling · Artificial intelligence · Life cycle assessment · Built environment · Climate change

R. Vergés (✉) · K. Gaspar · N. Forcada
Group of Construction Research and Innovation (GRIC), Universitat Politècnica de Catalunya, Terrassa, Catalonia, Spain
e-mail: roger.verges.eiras@upc.edu

D. Font
Eco Intelligent Growth, Sant Cugat del Vallès, Catalonia, Spain

A. Podleisek et al. (eds.), *Advances in Circular Economy: Bridging Research and Practice*, Springer Proceedings in Earth and Environmental Sciences,
https://doi.org/10.1007/978-3-032-21547-5_5

1 Introduction

The transformation of the built environment represents a quest for sustainable development, wherein the principles of the circular economy (CE) are often embraced as a necessary shift to address the current harmful, yet systemic, economic dynamics [25, 30], surpassing resource waste while emphasising conservation, reuse, and regeneration [18]. Even though these canons hold promise in the way global environmental and social concerns are overseen, they are not without their troubles in their basic conception [14, 28].

In stark contrast, CE is often considered a panacea for reducing carbon emissions, mitigating climate change, and improving overall efficiency. However, this technocratic engineered-centric perception can be misleading, as socio-cultural factors may undermine these anticipated benefits [32]. For instance, in William Stanley Jevons' work 'The Coal Question' [23], Jevon's Paradox is presented, suggesting that increased efficiency in a process consistently leads to a corresponding rise in demand for that process, negating, or even worsening, the initial improvements in performance. Hence, seeking an efficient solution without addressing the root causes of the problem constitutes a flawed approach, a principle that extends to the domain of CE as well. Frequently, these pathways are driven by self-interest motivations, prioritising economic accumulation as the ultimate goal [39]. As anticipated, the inclination toward avarice inherently undermines environmental sustainability, particularly given the finite nature of most resources. As a matter of fact, 'The Tragedy of the Commons' [20] illustrates the perils of overexploitation and degradation, especially in contexts where unrestricted access to finite resources is permitted. The challenges posed by the aforementioned threats remain pertinent even half a century later, as contemporary neoliberal economic paradigms repeatedly prioritise short-term profit over sustainable processes [17], even leading to green-washing behaviour practices [40]. This narrative caption the dire importance of responsible resource management and collaborative efforts to mitigate the adverse impacts of individualistic pursuits on shared environmental assets, even more in CE dynamics.

Over the decades, perceptions regarding the handling of CE have undergone various shifts. Illustrated in 'Steady-State Economics' [10], the prevailing academic consensus during the twentieth century advocated for achieving environmental sustainability by veering away from perpetual growth and embracing a steady-state approach instead. The aforementioned work, while influential in its time and onwards, it even faces challenges in meeting the sustainability imperatives of the present era. In the contemporary dynamic global economy, marked by rapid population expansion and escalating resource requirements, adopting a purely steady-state approach may be insufficient in confronting the intricate and evolving environmental and social challenges, especially without considering a prior phase of economic degrowth [9], necessary to be able to adjust to a sustainable, resilient, global scenario devoid of climate change and rebound effect [28]. As a result, the principles of CE offer a more adaptable and responsive framework, especially when considering circular degrowth [36], where waste is both reused and reduced. Moreover, recognising that the construction sector is indispensable due to perpetual human needs for inhabitable spaces that require construction, maintenance, refurbishment, and eventual demolition or recycling, it becomes imperative for the sector to transition towards a more sustainable framework, where CE must take the lead as a fundamental component within the built environment [31, 38].

To achieve these goals, the introduction of CE into the global economy is nearly essential, as just discussed. However, alongside implementation, it is inescapable to establish monitoring mechanisms to ensure its effective execution regarding sustainability objectives. Several studies have explored methodologies for measuring CE metrics, most of them expressing the need to use indicators with the aim of providing effective circular and sustainability information and tracking. For instance, González et al. [19] introduced a methodology framework aimed at delineating the degree of circularity within construction sector-oriented projects. However, this study, along with many others, introduce formulations that frequently lack interdependence and fail to provide a comprehensive view of the system's dynamics. Furthermore, the insights derived from academia have yet to fully permeate the construction sector, despite the emergence of certain policies and action plans geared towards promoting circularity [12]. One contributing factor to this gap is the tendency of many studies to focus on specific construction topics, such as the building phase, use, or demolition, rather than adopting a holistic perspective.

This oversight perpetuates an incomplete narrative, one that risks overlooking the intricate relationships and unintended consequences inherent in the pursuit of circularity. While the issue at hand is complex and systemic, incremental measures and small actions can be implemented to enhance overall sustainability within the construction sector. As a result, this paper aims to contribute by partially bridging this gap through the proposal of a holistic framework that promotes circularity in the construction sector to boost sustainability by considering its environmental, social, and economic aspects. To accomplish this, it is essential to assess the prevailing trends and demands in construction projects, extracting conclusions to attain a comprehensive understanding of sustainable construction, particularly within the framework of the circular economy.

This paper is structured as follows: the methods outline the research approach and methodologies employed, then an analysis of the prevailing trends and requirements is conducted. Subsequently, the framework proposal is presented, followed by an in-depth discussion. Finally, conclusions drawn from the findings are provided.

2 Methods

This research proposal is structured in two phases:

First, an analysis of prevailing trends and requirements is conducted to grasp the current landscape of construction projects, particularly within the contexts of the circular economy, sustainable development, and broader environmental, social, and economic considerations. This exploration encompasses an assessment of circularity and sustainability metrics currently in use, aiming to identify those with the potential to capture the most relevant interdependencies and environmental impacts. Also, regional, national, and European directives must be considered to understand the situation of sustainability and CE applied to construction.

Then, decision-making processes are evaluated to understand how construction projects operate and are managed. The examination assesses whether the involvement of experts introduces bias to the decision-making process, potentially hindering the implementation of circular strategies or sustainability practices. In relation to this, the integration of the decision-making process into the Building Information Modelling (BIM) environment, along with other automation and technological solutions, is explored. This aims to assess the potential for creating a comprehensive information environment and determine if enhancements are necessary or beneficial.

Second, utilising the gathered information, a comprehensive framework for circular sustainable construction is proposed, which encompasses the entire life cycle of the construction project. Special attention shall be given to achieving a holistic perspective, with a focus on waste reduction and recycling, specifically addressing circular degrowth. For that manner, Life Cycle Sustainability Assessment (LCSA) could serve as a driver of holisticness, as the whole life of a given product or material is taken in consideration with this procedure.

3 Prevailing Trends and Requirements

Three aspects in the context of prevailing trends and requirements within construction projects will be addressed: (i) Sustainability and circularity metrics within construction projects, (ii) Decision-making processes in product assessment, and (iii) Integration of BIM with complementary technologies and State-of-the-art holistic assessment of buildings throughout their entire life cycle.

3.1 Sustainability and Circularity Metrics Within Construction Projects

In understanding a construction project, it is essential to consider its diverse components. Buildings encompass structural elements such as roofs, facades, and walls, among others, constituting a significant portion of the project. Additionally, installation procedures play a crucial role, for instance involving plumbing, electrical, mechanical components, and integration of solar panels, among others. Furthermore, furniture, products, and last refinements contribute substantially to the overall design.

Each component comprises various materials, each offering distinct characteristics. Typically, there exist numerous options for a single component, each with its unique attributes. For example, facades may consist of brick, stone, concrete, metal, among others, with each alternative presenting specific costs, environmental impacts, social value, and potential for circularity. The selection of each product is frequently conducted without due consideration of environmental responsibility, as many companies prioritise profit maximisation [15]. For this reason, before considering the most suitable product for a given scenario, it is imperative to quantify specific metrics that address not only economic aspects but also environmental and social concerns.

As asserted by Raworth [33], nine planetary boundaries pose threats to well-being and sustainability, encompassing critical factors such as climate change, biodiversity loss, and land conversion. Additionally, the social aspects, intertwined with these ecological considerations, highlight the interdependencies between human welfare and environmental health.

Similarly, one approach to address and evaluate these concerns is to focus on the product life cycle via certifications and regulations, which can be regional or international. Examples include UNE-EN ISO 14040 [6], UNE-EN ISO 14044 [4], UNE-EN 15804 [5], and UNE-EN 15978 [3]. From these regulations, Environmental Product Declarations (EPDs) can be generated in a standardised manner, enabling consistent comparison between construction items, if these are based on the same Product Category Rules (PCRs). At the European level, the Level(s) [13] framework provides a foundation of practices and indicators for assessing sustainability in the construction sector and, thus, the proposed solution in this paper is also in alignment with these frameworks.

Also, one of the most significant considerations in evaluating the environmental, social, and economic impacts of a construction project is the site-specific dimension of the associated metrics and information, with the aim of obtaining more nuanced, reliable results [34].

EPDs have common environmental information. Table 1 summarises them. Should location-specific databases be used depending on the construction project, along with the information provided by the EPD, comparable results could be extracted in the context of the framework proposal.

As can be seen, these metrics enable the evaluation of environmental impacts associated with the various products present within a construction project.

Nevertheless, the subsequent circularity analysis of these metrics and the evaluation of sustainability assessments lack standardisation. Therefore, it is imperative to establish common mechanisms of comparison to assess the most suitable item, even if none of the products are perfectly suitable in all aspects. For example, a lower global warming potential indicates a more favourable option in terms of greenhouse gas (GHG) emissions, increasing the likelihood of selecting that product compared to its alternatives, even though perhaps none of these options is environmentally friendly. The waste and raw material data also enable the quantification of the circularity associated with each specific element by evaluating the extent to which materials are reused, recycled, or recovered within the system.

Table 1 Available information in databases using EPDs

Item	Unit	Impact categories
Normalised costs	€ / kg (or m^3)	–
Global warming potential (CO_2 *eq*)	kg	–
Primary energy consumption	MJ	Renewable and non-renewable primary energy
Impact on the environment	m^3, kg, MJ	Stratospheric ozone depletion potential, acidification potential of soil and water resources, eutrophication potential, and tropospheric ozone formation potential
Consumption of natural resources	m^3, MJ, kg	Water consumption, depletion potential of fossil abiotic resources, and depletion potential of non-fossil abiotic resources
Raw material	%	–
Waste (if deemed applicable)	kg	Placement waste, selective separation, minimum selective separation by waste type

3.2 Decision-Making Processes in Product Assessment

In construction projects, Multiple Criteria Decision Analysis (MCDA) is widely regarded as the most common practice for assessing the best alternative among various products, items, or materials, due to the existence of conflicting objectives (e.g., a sustainable alternative to a conventional product is more expensive and thus conflictive in brand economic objectives) [41]. For this reason, the decision-maker typically plays a primary role, as they are responsible for evaluating these discrepancies and ultimately selecting the most suitable option.

In fact, MCDA is often conducted without the support of decision support systems, which are rare in construction projects due to the challenges associated with their development, primarily stemming from the scarcity of available data [29]. Should these systems be implemented, they could be powered by AI, employing machine learning technologies such as artificial neural networks for the selection of suitable alternatives and leveraging Bayesian networks to address and overcome uncertainties in the decision-making process [26].

As just mentioned, academia advancements lean towards employing expert-enhanced decision-making criteria to determine the most appropriate construction item [16]. Although this approach is valuable on its own, and results tend to be consistent and robust, it is also important to note that each professional harbours their unique perspective and may prioritise one metric over another [7]. So, human decision-making is inherently biassed, thus rendering it highly improbable for a formula to effortlessly capture and predict this process.

Depending on stakeholder prioritisation, a variety of expert types may be consulted. However, this approach inherently possesses flaws and biases. Conversely, it would be advantageous to seek a broader opinion incorporating diverse experts to mitigate personal inconsistencies and opinions regarding each specific metric. Nevertheless, conducting such an exhaustive task each time a product typology needs to be selected is deemed impractical, time-consuming, and financially burdensome.

An alternative solution arises in the aforementioned AI-related systems, wherein all these inputs can be aggregated to train a model serving as a nearly neutral expert to discern the most suitable construction item, without overlooking rarer metrics, and always prioritising sustainability by fostering circularity. However, it is essential that the training and testing data be consistent, robust, and comprehensive, often a difficult approach, as discussed above.

3.3 Integration of BIM with Complementary Technologies and Holistic Overview

BIM is a collaborative methodology that facilitates the handling, processing, and accessibility of all building-related information throughout its entire life cycle, with end-of-life information being uncommon, but feasible and increasingly being implemented [37]. With BIM, traceability of materials is enabled, allowing for the assessment of environmental, social, and economic impacts. However, two significant issues regarding BIM persist: its implementation in businesses and its integration with complementary technologies.

Despite the tangible advancements facilitated by BIM, several barriers impede the full implementation of the methodology within firms. These include social and habitual resistance to change, reliance on traditional methods of contracting, high expenses associated with training and the steep learning curve, costly software procurement, and a general lack of awareness about BIM [1]. Fortunately, directives around the globe have begun to promote the use of BIM in construction projects since the 2010s. For instance, Directive 2014/24/EU [11] serves as an example of such promotion within the European context, contributing to the solid adoption of BIM in these areas [8]. However, developing countries still require time to adopt BIM on a widespread scale [2].

The utilisation of BIM can expedite the process of evaluating the environmental impact of various product configurations—or even the services related to the building—and understanding their interdependencies. The assessment of this approach is conducted externally to the BIM environment; however, it uses BIM data, such as the total weight of a material, the number of certain items, among other relevant information. Thus, the integration of BIM with complementary technologies such as AI-enhanced product selectors or LCSA software emerges as the most feasible option.

As discussed in the introduction, the construction sector lacks documented holistic overviews of sustainability. This gap is significant because incomplete assessments may lead to erroneous conclusions if only certain stages of the building process are considered. Holistic assessment of environmental, social, and economic impacts of buildings is formed through the consideration of the life cycle assessment of the project, spanning from its initial product stage to its eventual end of life, following the cradle-to-cradle methodology outlined in ISO 14040/44 [21].

BIM has become a prominent tool in recent research, primarily employed to extract quantities essential for establishing the Life Cycle Inventory (LCI) needed for LCSA. However, challenges arise when conducting LCSAs during early design stages using BIM, as the models often lack the necessary level of geometry and information for a comprehensive assessment. Consequently, BIM-based evaluations at early stages require significant manual input, resulting in increased complexity and time consumption. As a result, many BIM-LCSA studies opt for one-time retrospective assessments [27]. While BIM-based assessments effectively document environmental performance during detailed design and construction phases, they offer limited potential for large-scale design improvements. Given these limitations, data visualisation emerges as a promising alternative to address these constraints [35]. With this approach, it becomes feasible to explore various alternatives based on product selection by visualising the environmental, social, and economic impacts of each option. Consequently, iterative processes may be both necessary and advisable to attain the most optimal solution. However, data visualisation within the context of the BIM environment remains limited and underdeveloped, lacking systematic implementation [22].

4 Framework Proposal and Discussion

In light of the prevailing landscape surrounding circularity, LCSA, BIM, and integrated technologies within the building sector, this section introduces a framework proposal as an initial step toward addressing existing flaws, unexplored areas, and unresolved issues in the field. Before presenting the framework, it is essential to outline the conclusions drawn from each analysis within the sector conducted previously.

First, circularity building metrics are not frequently integrated into LCSA processes due to differing objectives. Whereas LCSA primarily evaluates the environmental or sustainability aspects of a product, circularity metrics, particularly those considering circular degrowth, offer a fresh perspective emphasising waste reduction and optimisation through both reduction and reuse strategies. However, it should be noted that the ultimate goal is to measure environmental impacts. Circularity metrics are introduced as an intermediate step to provide a deeper understanding of the system and to determine if CE practices are suitable for each specific case.

Second, automated decision-making criteria are not commonly employed in construction projects, as the decision-making process typically relies on human experts, which may introduce bias. AI-based models, if correctly trained—i.e., not only prioritising economic objectives as most companies would do-, present an opportunity to introduce more neutrality while optimising economic resource allocation and time efficiency.

Third, data visualisation is not yet a common practice in BIM environments, resulting in significant information being overlooked. Should visualisation tools be employed, there would be an opportunity for holistic assessment and iterative selection of materials, products, and items. This approach allows for the minimization and mitigation of global impacts to the greatest extent possible.

Fourth, product selection should not occur in isolation; rather, it needs to consider the entire family group within the project, such as facades, walls, and other components. Synergies exist among these elements, and their mutual interactions can significantly impact the sustainability assessment of the building design [24]. As a reaction, system typologies should be evaluated instead. Consequently, the smart AI-based selector must integrate the entire project scope to yield optimal results for any given combination. Thus, considering these factors, the resulting framework proposal is as depicted in Fig. 1.

In this framework, a product database, coupled with its quantities concretisation using BIM, will form the cornerstone of preliminary building project information. This data will then be fed into the software, which will conduct the LCSA with consideration of the location-specific properties, leading to more precise outcomes. Subsequently, the smart selector will identify the most suitable option, particularly in cases where conflicting priorities arise, such as economic versus environmental concerns. Additionally, and after conducting data visualisation procedures (whether typical software and/or VR) final calculations related to circular economy and circular degrowth will be performed to achieve a holistic overview. The objective would be to increase sustainability by maximising circular recycling while minimising waste in any form, in those cases deemed appropriate. If necessary, the process can be iterated with additional alternatives for further optimisation and to raise awareness of critical choke points or areas to avoid.

The inputs of the machine learning model would comprise matrices representing the indicators defining the sustainability, circularity, and costs of a construction item. Each input matrix contains the metrics associated with each alternative.

The output should consist of a sorted list representing the suitability of each item in comparison to the others. The raw output matrix is a numerical vector with the same number of rows as items, with each value corresponding to the sorted list presented in the input, i.e., the strength of choice or preference score. The overall neural network proposed scheme is illustrated in Fig. 2.

5 Conclusions

The analysis of current trends and requirements within construction projects regarding CE and sustainability allows for the development of a holistic framework proposal. This framework aims to address the challenges and inconsistencies present in the building design process as it stands today. The following conclusions can be extracted:

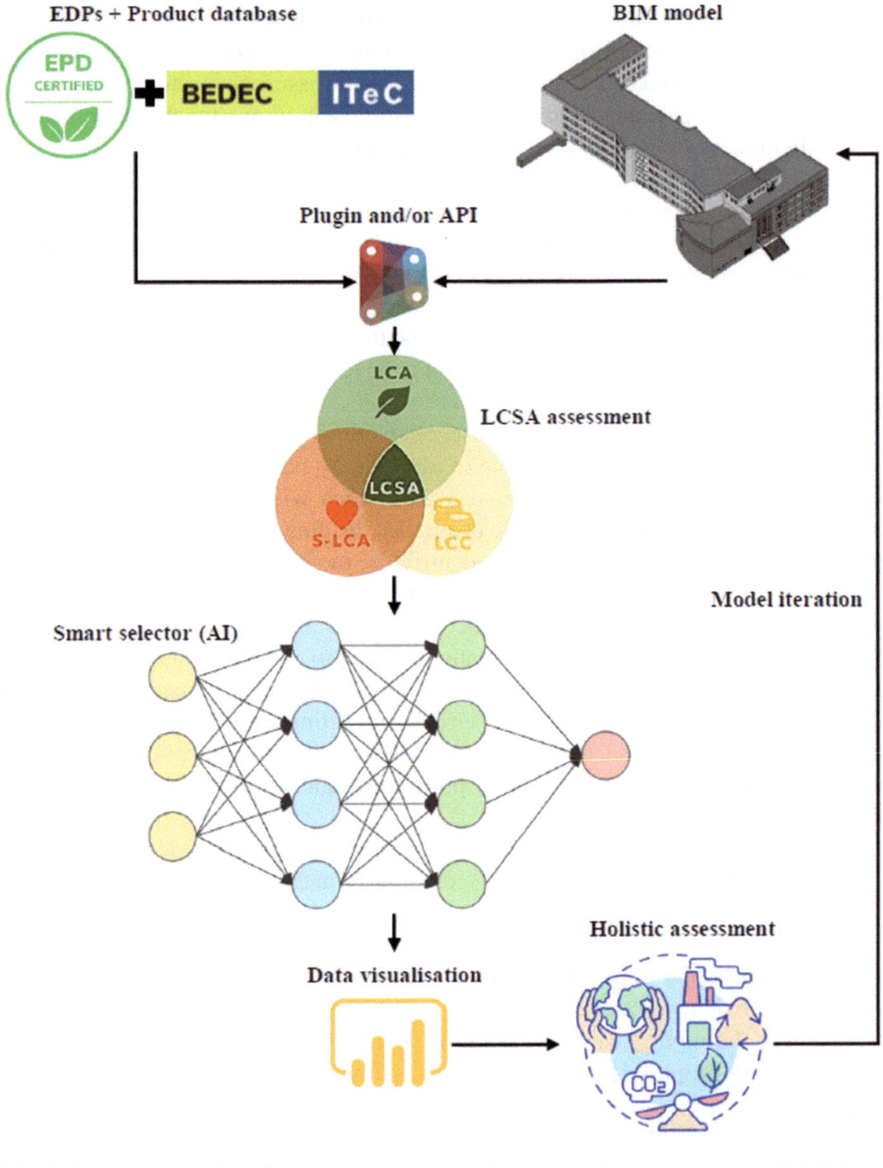

Fig. 1 Overview of the holistic assessment procedure for construction projects using AI, BIM, and sustainability and circularity metrics

- Circularity metrics are seldom incorporated into LCSA processes due to their distinct objectives in nature. These metrics could provide intermediate information with the aim of providing a deeper understanding of the system and to determine if CE practices are suitable for each specific case.

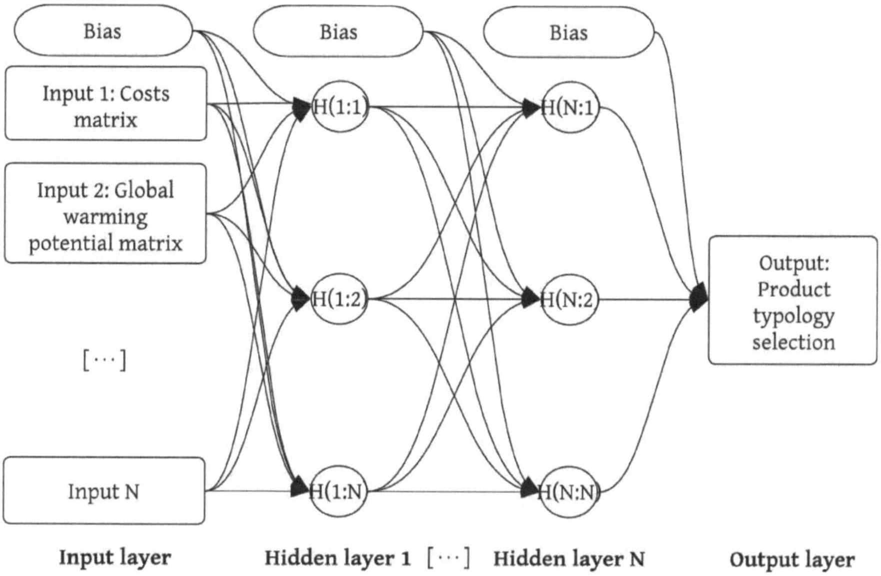

Fig. 2 Schematic of the decision criteria for construction typologies based on AI

- Automated decision-making criteria are infrequently utilised in construction projects, where decisions often rely on human experts, introducing potential biases.
- Data visualisation remains uncommon in BIM environments, leading to substantial oversight of valuable information.
- Product selection should not be conducted in isolation but rather should consider the entire family group within the project, including facades, walls, and other components, i.e., system typologies. Synergies among these elements can have a significant impact on the sustainability assessment of the building design, specifically when considering all the life cycle phases of the building.

Acknowledgement This research was supported by the Catalan agency AGAUR through its Industrial Doctorate support program (2023 DI 00037). The authors would like to express their gratitude to *Grupo Construcía* for their support in executing this research and for providing the necessary data to conduct it.

References

1. Ahmed, S. (2018). Barriers to Implementation of Building Information Modeling (BIM) to the Construction Industry: A Review. Journal of Civil Engineering and Construction, 7(2), 107–113. https://doi.org/10.32732/JCEC.2018.7.2.107

2. Ariono, B., Wasesa, M., & Dhewanto, W. (2022). The Drivers, Barriers, and Enablers of Building Information Modeling (BIM) Innovation in Developing Countries: Insights from Systematic Literature Review and Comparative Analysis. Buildings 2022, Vol. 12, Page 1912, 12(11), 1912. https://doi.org/10.3390/BUILDINGS12111912
3. Asociación Española de Normalización (2012). UNE-EN 15978:2012. Sustainability of construction works—Assessment of environmental performance of buildings—Calculation method. Available at: https://www.une.org/encuentra-tu-norma/busca-tu-norma/norma?c=N0049397
4. Asociación Española de Normalización (2018). UNE-EN ISO 14044:2006/A1:2018. Environmental management—Life cycle assessment—Requirements and guidelines—Amendment 1 (ISO 14044:2006/Amd 1:2017). Available at: https://www.une.org/encuentra-tu-norma/busca-tu-norma/norma?c=N0060852
5. Asociación Española de Normalización (2020). UNE-EN 15804:2012+A2:2020. Sustainability of construction works—Environmental product declarations—Core rules for the product category of construction products. Available at: https://www.une.org/encuentra-tu-norma/busca-tu-norma/norma?c=N0063508
6. Asociación Española de Normalización (2022). UNE-EN ISO 14040:2006/A1:2021. Environmental management—Life cycle assessment—Principles and framework—Amendment 1 (ISO 14040:2006/Amd 1:2020). Available at: https://www.en.une.org/encuentra-tu-norma/busca-tu-norma/norma?c=N0066670
7. Božiček, D., Almezeraani, Y., & Košir, M. (2023). Making sense of LCA results when evaluating multiple building designs–comparison of interpretation concepts. Building Research and Information. https://doi.org/10.1080/09613218.2023.2236254
8. Charef, R., Emmitt, S., Alaka, H., & Fouchal, F. (2019). Building Information Modelling adoption in the European Union: An overview. Journal of Building Engineering, 25, 100777. https://doi.org/10.1016/J.JOBE.2019.100777
9. Czech, B., & Daly, H. E. (2014). The steady state economy as the sustainable alternative to economic growth. Peak Oil, Economic Growth, and Wildlife Conservation, 119–129. https://doi.org/10.1007/978-1-4939-1954-3_6/COVER
10. Daly, H. E. (1977). Steady-state economics: the economics of biophysical equilibrium and moral growth. 185
11. European Commission (2014). Directive 2014/24/EU of the European Parliament and of the Council of 26 February 2014 on public procurement and repealing Directive 2004/18/EC (Text with EEA relevance)Text with EEA relevance. Available at: https://eur-lex.europa.eu/legal-content/EN/TXT/?uri=CELEX%3A02014L0024-20240101
12. European Commission (2020). COMMUNICATION FROM THE COMMISSION TO THE EUROPEAN PARLIAMENT, THE COUNCIL, THE EUROPEAN ECONOMIC AND SOCIAL COMMITTEE AND THE COMMITTEE OF THE REGIONS A new Circular Economy Action Plan For a cleaner and more competitive Europe (COM/2020/98). Available at: https://eur-lex.europa.eu/legal-content/EN/TXT/?qid=1583933814386&uri=COM:2020:98:FIN
13. European Commission (2022). Level(s) European framework for sustainable buildings. Available at: https://environment.ec.europa.eu/topics/circular-economy/levels_en
14. Figge, F., & Thorpe, A. S. (2019). The symbiotic rebound effect in the circular economy. Ecological Economics, 163, 61–69. https://doi.org/10.1016/J.ECOLECON.2019.04.028
15. Gabel, H.; Sinclair-Desgagné B. (1998). The firm, its routines, and the environment.
16. Figueiredo, K., Pierott, R., Hammad, A. W. A., & Haddad, A. (2021). Sustainable material choice for construction projects: A Life Cycle Sustainability Assessment framework based on BIM and Fuzzy-AHP. Building and Environment, 196, 107805. https://doi.org/10.1016/J.BUILDENV.2021.107805
17. Garrity, E. J. (2012). Tragedy of the Commons, Business Growth and the Fundamental Sustainability Problem. Sustainability 2012, Vol. 4, Pages 2443-2471, 4(10), 2443–2471. https://doi.org/10.3390/SU4102443

18. Ghisellini, P., Ripa, M., & Ulgiati, S. (2018). Exploring environmental and economic costs and benefits of a circular economy approach to the construction and demolition sector. A literature review. Journal of Cleaner Production, 178, 618–643. https://doi.org/10.1016/J.JCLEPRO.2017.11.207
19. González, A., Sendra, C., Herena, A., Rosquillas, M., & Vaz, D. (2021). Methodology to assess the circularity in building construction and refurbishment activities. Resources, Conservation & Recycling Advances, 12, 200051. https://doi.org/10.1016/J.RCRADV.2021.200051
20. Hardin, G. (1968). The Tragedy of the Commons. Science, 162(3859), 1243–1248. https://doi.org/10.1126/SCIENCE.162.3859.1243
21. International Organization for Standardization (ISO). (2006). ISO 14040:2006. Environmental management. Life cycle assessment. Available at: https://www.iso.org/standard/37456.html
22. Ivson, P., Moreira, A., Queiroz, F., Santos, W., & Celes, W. (2020). A Systematic Review of Visualization in Building Information Modeling. IEEE Transactions on Visualization and Computer Graphics, 26(10), 3109–3127. https://doi.org/10.1109/TVCG.2019.2907583
23. Jevons, W. (1865). The Coal Question; an Inquiry concerning the Progress of the Nation, and the Probable Exhaustion of our Coal Mines. Macmillan & Co. https://doi.org/10.1017/S0016756800197468
24. Ju, C., Ning, Y., & Pan, W. (2016). A review of interdependence of sustainable building. Environmental Impact Assessment Review, 56, 120–127. https://doi.org/10.1016/J.EIAR.2015.09.006
25. Khan, M., Ullah, K. I. A., Nasir, F., al Alahmadi, A. R., Alzaed, A. A., Alwetaishi, A. N., Ghufran, M., Iqbal, K., Khan, A., Ullah, F., Nasir, A. R., al Alahmadi, A. A., Alzaed, A. N., & Alwetaishi, M. (2022). Circular Economy in the Construction Industry: A Step towards Sustainable Development. Buildings 2022, Vol. 12, Page 1004, 12(7), 1004. https://doi.org/10.3390/BUILDINGS12071004
26. Kjærulff, U. B., & Madsen, A. L. (2008). Bayesian Networks and Influence Diagrams: A Guide to Construction and Analysis. https://doi.org/10.1007/978-0-387-74101-7
27. Llatas, C., Soust-Verdaguer, B., Hollberg, A., Palumbo, E., & Quiñones, R. (2022). BIM-based LCSA application in early design stages using IFC. Automation in Construction, 138, 104259. https://doi.org/10.1016/J.AUTCON.2022.104259
28. Makov, T., & Font, D. (2018). Does the circular economy grow the pie? The case of rebound effects from smartphone reuse. Frontiers in Energy Research, 6(MAY), 368568. https://doi.org/10.3389/FENRG.2018.00039/BIBTEX
29. Marcher, C., Giusti, A., & Matt, D. T. (2020). Decision Support in Building Construction: A Systematic Review of Methods and Application Areas. Buildings 2020, Vol. 10, Page 170, 10(10), 170. https://doi.org/10.3390/BUILDINGS10100170
30. Morseletto, P. (2023). Sometimes linear, sometimes circular: States of the economy and transitions to the future. Journal of Cleaner Production, 390, 136138. https://doi.org/10.1016/J.JCLEPRO.2023.136138
31. Oluleye, B. I., Chan, D. W. M., Saka, A. B., & Olawumi, T. O. (2022). Circular economy research on building construction and demolition waste: A review of current trends and future research directions. Journal of Cleaner Production, 357, 131927. https://doi.org/10.1016/J.JCLEPRO.2022.131927
32. Polimeni, J. M., Mayumi, K., Giampietro, M., & Alcott, B. (2012). The Jevons Paradox and the Myth of Resource Efficiency Improvements. Earthscan, 1–184. https://doi.org/10.4324/9781849773102
33. Raworth, K. (2017). A Doughnut for the Anthropocene: humanity's compass in the 21st century. The Lancet Planetary Health, 1(2), e48–e49. https://doi.org/10.1016/S2542-5196(17)30028-1
34. Reap, J., Bras, B., Newcomb, P. J., & Carmichael, C. (2008). Improving Life Cycle Assessment by Including Spatial, Dynamic and Place-Based Modeling. Proceedings of the ASME Design Engineering Technical Conference, 3, 77–83. https://doi.org/10.1115/DETC2003/DFM-48140
35. Röck, M., Hollberg, A., Habert, G., & Passer, A. (2018). LCA and BIM: Visualization of environmental potentials in building construction at early design stages. Building and Environment, 140, 153–161. https://doi.org/10.1016/J.BUILDENV.2018.05.006

36. Savini, F. (2023). Futures of the social metabolism: Degrowth, circular economy and the value of waste. Futures, 150, 103180. https://doi.org/10.1016/J.FUTURES.2023.103180
37. Shi, Y., & Xu, J. (2021). BIM-based information system for econo-enviro-friendly end-of-life disposal of construction and demolition waste. Automation in Construction, 125, 103611. https://doi.org/10.1016/J.AUTCON.2021.103611
38. Stephan, A., & Athanassiadis, A. (2018). Towards a more circular construction sector: Estimating and spatialising current and future non-structural material replacement flows to maintain urban building stocks. Resources, Conservation and Recycling, 129, 248–262. https://doi.org/10.1016/J.RESCONREC.2017.09.022
39. Verburg, R. (2018). Greed, self-interest and the shaping of economics. Taylor & Francis, 1–216. https://doi.org/10.4324/9781315269023/GREED-SELF-INTEREST-SHAPING-ECONOMICS-RUDI-VERBURG
40. Vieira de Freitas, S., Sobral, M. F. F., Ribeiro, A. R. B., & Soares, G. R. da L. (2020). Concepts and forms of greenwashing: a systematic review. Environmental Sciences Europe, 32(1), 1–12. https://doi.org/10.1186/S12302-020-0300-3/FIGURES/5
41. Zavadskas, E. K., Antucheviciene, J., Vilutiene, T., & Adeli, H. (2017). Sustainable Decision-Making in Civil Engineering, Construction and Building Technology. Sustainability 2018, Vol. 10, Page 14, 10(1), 14. https://doi.org/10.3390/SU10010014

Mycelium Material Substitutes Conventional Bicycle Helmet Material by Adapted Design for X Processes

Salome Berger, Simon Winterberg, Corinna Baumgartner, Jan Inauen, Justin Staller, and Daniel Schmid

Abstract Evolving policy frameworks and social pressure for more sustainable circular products stipulate new requirements for the product development process. Thus, the Design for X (DfX) processes are adapted for these requirements and applied to a case study that aims to transition a conventional bicycle helmet to a fully circular product that adheres to the Cradle to Cradle principle with a biological cycle. The DfX processes and the resulting considerations are utilized to investigate and overcome the challenges posed by the application of biological materials, such as mycelium, to a massproduced B2C product. Thereby the concept for a suitable composite comprising of several organic materials is conceived and tested. A new manufacturing process capable of massproducing helmets with said composite is developed with the DfX processes to make the concept economically viable. Initial drop-tests of helmet prototypes prove that the redesigned helmets have similar protective damping features to conventional helmets, and thus are functionally viable. The successful development of the bicycle helmet case provides evidence that the adaptation of the holistic DfX approach is feasible for overcoming the challenge of product development processes with increasingly complex new requirements.

Keywords Cradle to Cradle · Design for X · Bioeconomy · Product development

S. Berger (✉) · S. Winterberg · C. Baumgartner · J. Inauen · J. Staller · D. Schmid
Zurich University of Applied Sciences, Winterthur, Switzerland
e-mail: begr@zhaw.ch

S. Winterberg
e-mail: wins@zhaw.ch

C. Baumgartner
e-mail: bamo@zhaw.ch

J. Inauen
e-mail: inau@zhaw.ch

J. Staller
e-mail: xsju@zhaw.ch

D. Schmid
e-mail: scdd@zhaw.ch

A. Podleisek et al. (eds.), *Advances in Circular Economy: Bridging Research and Practice*, Springer Proceedings in Earth and Environmental Sciences,
https://doi.org/10.1007/978-3-032-21547-5_6

1 Introduction

1.1 Impact of the Circular Economy

Evolving policy framework. The policy framework for new products is evolving significantly. They must fulfil sustainability requirements, ideally according to the Cradle to Cradle principle. The European Union (EU) launched the European Green Deal [1] in 2019, striving to be the first climate-neutral continent, and published "Industry 5.0, a transformative vision for Europe" [2] in 2022, where sustainability is one of the three pillars. The direction is set by the EU's circular economy action plan [3], which transforms the legal framework of the economy and affects Switzerland as the EU is its most significant economic partner. 50% of the goods exported from Switzerland are exported to the EU (2021: 130.26 Billion CHF, 50.14%) [4]. The MEM industry's share in exported goods is 27% [5], while the destination of 29% is the EU [6]. Therefore, product developments must fulfil those sustainability standards in the future.

Product benefits. Product benefits consist of basic benefits, additional benefits (environment), edification benefits (aesthetic aspects), and prestige benefits (social effect) [7, p. 396]. The benefits of new products that fullfil the requirements of the circular economy, here with a clear focus on the biological cycle [8, 9] and not on the technical, deviate from product benefits as they are known today. New materials result, e.g. in new forms, shapes, and haptics. The customer experience and the customer perceptions have to change. Furthermore, the product design process is evolving as new materials affect the product's manufacturing, assembly, testing, maintenance, reliability, and verification to reduce the environmental impact.

1.2 Necessity for Adapted Design for X Processes

Product developments are highly interactive, interdisciplinary, and require iterative loops [10]. If a new material, in this case fulfilling the Cradle to Cradle principle, has to be applied, many aspects of the product development change [11] and new compromises and optimums need to be found. Design for X (DfX) summarises the following focus areas which have to be considered [12, Ch. 1.6]: Assembly (DfA), manufacturing (DfM), test and maintenance (DfTM), verification (DfV), reliability (DfR), and environment (DfE).

The adaption of the DfX processes shall enable the substitution of a fossile-based material with biological mycelium material. The processes are sorted into four dimensions: physical, design, production, and environmental:

- *Physical dimension.* Damping property, strength, de-mouldability, ultraviolet (UV) radiation resistance/durability, temperature resistance, moisture and water

resistance and material behaviour (e.g. against scratches and minor impacts) of the mycelium material
- *Design dimension.* Design of the helmet (e.g. favourability and aesthetics), comfort of the helmet (shape, ventilation and weight), haptic (interior and exterior material surface)
- *Production dimension.* Manufacturing (shaping, moulding, filling), assembly (parts integration, joining technology), automatization (serial production)
- *Environmental dimension.* Environmental life-cycle performance.

Further considerations that are excluded from the early stage development are material behaviour in daily usage (e.g. against scratches and minor impacts), durability of the product and material, administration, sales, and distribution costs.

Research question. The following research question was formulated to analyze what impact following circular economy requirements has on product development: *How must Design for X (DfX) be adapted so that biological mycelium material can replace foamed plastic granulate (poly-styrene, ESP) and a combined plastic shell element by using in-mould technology and fulfilling the Cradle to Cradle principle?*

Selected product example—bicycle helmet. Bicycle helmets are essential for cyclists to protect themselves [13]. It is a purely mechanical serial product in the B2C market and of limited complexity. The helmet has different components [14], e.g. the shell and the expanded polystyrene (EPS) foam, which are either glued together ("softshell") or welded together ("hardshell") [15].

Today, those components are mostly based on fossil materials. Therefore, using the bicycle helmet as a case study for adapting the product development via the DfX processes and realizing the transfer into the biological circle of the Cradle to Cradle principle is ideal. Furthermore, the new helmet with the mycelium material fulfils the requirements of the Circular Bioeconomy [16–18].

2 Methods—DfX

As described in Sect. 1, "Design for X" (DfX) contains various aspects of how products must be designed. In this specific case, the methods entail the following design processes:

Developing design guidelines (DfM). Mycelium-based products have specific design features that differ from classic plastic injection moulded products [19]. Thus, the crucial features are engineered in iterative steps based on research and physical tests to derive related parameters for the demolding angle, the wall thickness, the material radii, and the surface roughness. These parameters are interconnected and affect other DfX dimensions, e.g. the appearance and haptic.

Testing material properties (DfR, DfTM). The resistance of mycelium to rain, UV radiation, dirt, and abrasion is insufficient for use as a bicycle helmet, and its appearance deviates from today's material (aesthetics, surface roughness). Therefore, the disadvantages have to be compensated with at least one protective layer, achieving a resistance to the environmental conditions mentioned above [20, 21]. Testing different samples with different protective layers is the method to define the rules for DfR and DfTM.

Developing production guidelines (DfM). As no product is without production; new materials must be understood to conceive and evaluate potential production processes. Subsequently, exploratory research uses the qualitative methodological approach and iterative improvement loops to create insights for the bio-based mycelium composite [19]. To compare the production setups, the suitability of the processes is analyzed and evaluated, including the number and type of process steps involved, as well as the tools and materials used, while considering the specific product characteristics, quality, and efficiency. Strengths and weaknesses of the processes are identified and the potential for improvement is evaluated. The analysis also includes the consideration of scalability and the potential for industrialization of the process based on factors such as automation and standardization. This comparison enables the transition from prototype testing to the industrial manufacturing process.

Designing Assembly Guidelines (DfA). In general, assembly guidelines are based on designing on how single elements fit together. It cannot be done independently of the product itself. A bicycle helmet has three main components: the helmet body or liner (the central part of a helmet with shock absorption as the primary function), an outer shell, and a fastening system. These elements must be combined during the assembly process.

For the design of the assembly specifications, different procedures are defined and divided into two groups: connecting the helmet shell and the protective layer (outer layer assembly) and fixing the strap system (fastening assembly).

Existing principles for the connection of the protective layer and helmet shell are examined, evaluated, and experimentally tested. When attaching the bracket, the focus is on the adhesion properties of the mycelium [22]. Various methods for joining inserts from classic plastic production and existing solutions for conventional helmets must be analyzed and adapted to the mycelium conditions through sample tests.

Design for Environment (DfE). The transition to a biological material and the cradle to cradle approach ensures environmental sustainability. To verify this claim, a comparative fast-track life cycle assessment (LCA) of the environmental performance of different helmet variants, including a conventional helmet, is conducted.

The functional unit of the LCA is a bicycle helmet over its entire life cycle from cradle to grave without the utilization phase but including credits for recycling. The comparison is carried out using two different life cycle impact assessment methods with SimaPro 9.2 and the ecoinvent database 3.7 [23, 24]:

- *Ecological Scarcity* 2013 V1.06 in ecopoints: Level characterization with accumulated midpoint indicators [25]
- Global warming potential (GWP) using the method *IPCC GWP 100a (2013) V 1.03* in kg CO_2-eq [26].

Applying the Test Standards (DfV). Bicycle helmets, mycelium and commercially available ones, are drop tested according to the standard DIN EN 1078:2014-04 [27] to provide proof of concept for the protective function. The maximum acceleration value must not exceed 250 g. Mycelium samples are tested at each stage of the development process. Temperature and ageing influences, as well as the tests regarding field of view and fixation on the head have not been taken into account.

2.1 Physical Properties

The general physical properties of mycelium composites are known and described in "Engineered mycelium composite construction materials from fungal biorefineries: A critical review" [24]. The bicycle helmet prototypes and samples were all made of a preincubated mixture (gravimetric proportions) [28]:

- 46% Composition Substrate (dry): 49.9% hemp stalk chips, 49.9% (50/50 mix of spruce & fir sawdust), 0.2% $Ca(OH)_2$, incubated with strain: Trametes Versicolor
- 54% water.

As with conventional materials, the de-mouldability of mycelium composites depends on the draft angles, minimum radii, and the surface quality of the growth mould, which was demonstrated in the first tests with mycelium helmet prototypes. Due to unoptimized growth moulds, initial de-mouldings were difficult and led to cracked structures in the helmet (Fig. 1). Thus, the growth mould was improved with wider bars (increased from min. width of 17 to 31 mm), increased pull-out bevels (increased from 4.5° to 7°), larger minimum radii (increased from radius 1 to 3 mm) and a smoother surface (improved from average Rz of 97 μm to Rz of 85 μm). Tests conducted with the optimized model and helmet design de-moulded without significant problems, so no further optimization was required. Two external manufacturers of mycelium products grow bicycle helmets using their processes and mycelium mixtures in the optimized moulds for verification purposes. The results confirm the proof of concept. The filling, growing and de-moulding process were executed without any problems.

2.2 Design and Product Properties

Design and acceptance. The geometrical design of the bicycle helmet is conceived to resemble a conventional helmet to facilitate a direct comparison.

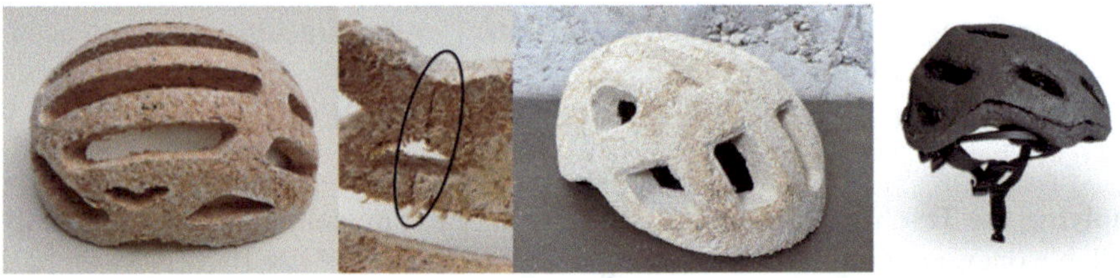

Fig. 1 Engineering loops of mycelium helmet (from left to right) with de-moulding issues of first prototype (second from left)

The use of mycelium harbours new challenges in aesthetics and design, because a bicycle helmet is a mass product that is primarily differentiated and sold through design aspects. Mycelium is an organically grown structure that has inconsistent visual characteristics. Compared to a conventional helmet, the surface is rough and inhomogeneous, forming a porous, uneven surface with shadows and colour changes.

An essential aspect of the anticipated market introduction of such a bicycle helmet is the customers' acceptance of it. This could pose a significant challenges to mycelium, especially because a helmet is worn on the head in direct contact to the consumer. Users might associate its structure and surface properties with common fungi and trigger rejection or even disgust. In this regard, utilizing a protective layer made of paper, similar to cardboard (pulp shell) is advantageous. It increases the helmet's chances of success by concealing the mycelium structure. The paper layer has a familiar, pleasant, and smooth haptic. Today's industrially manufactured pulp products can be produced with a high visual and haptic homogeneity and meet the customer's design requirements better.

The weight of commercially available helmets, an essential parameter for comfort, lies between 200 and 350 g. With a weight of 284 g, the developed helmet can keep up with conventional helmets.

Pulp shell as a protective layer. To increase the helmet's resistance towards moisture and UV, various coating concepts were considered, including PLA (polylactic acid), flax-gelatine and flax R*Concept Epoxy composite.

The LCA (see 2.4) demonstrates that plant-based protective shells are preferred, due to their lower environmental impact. This led us to the idea of using biodegradable paper. As a plant-based material for the helmet shell, paper has the advantage of easy decomposition when compared to PLA, which is only fast and fully compostable under certain conditions (in highly basic and highly acidic mediums) [29]. Thus, to produce the helmet's outer hard shell, wastepaper was shredded, mixed with water and stirred. A commercially available solution (Bioseal from Impershield) was also tested in combination with the paper pulp. The product, which consists of tree resin and plant fibres, was specially developed to protect bio-based materials by providing high water resistance and compostability at the same time. The combination of a mycelium liner and pulp shell is very suitable for an industrial manufacturing process.

The pulp hard shell is an ideal mould for the mycelium, which can grow into the shell. The fact that mycelium can fuse with paper pulp solves the bonding problems.

In combination with the above mentioned design consideration, this leads to the decision to further develop the protoyp helm with an outer and inner shell made of paper pulp.

2.3 Production Dimension

Using the established production processes, mycelium can be produced in a repeatable manner with adherence to the parameters known from the literature: hygienic environment and equipment, sufficient carbohydrates, adjusted humidity level, ambient temperature and a growth time of 5 days. A growth period of 3 days in the mould, followed by 2 days without mould and an increased ambient humidity of at least 95% rH significantly increases the formation of a mycelium outer skin. The demouldability depends on the factors described in 2.1.

The long in-mould growth process of the mycelium results in a large quantity of necessary moulds per product and time. Such moulds would either have to be one-way moulds, defeating the sustainability and circularity aspect of the product, or expensive multi-use moulds that are blocked for a long time, making an economical production process comparable to synthetic foams impossible. The newly conceived approach eliminates this disadvantage, by letting the mycelium grow directly into a massproduced pulp shell, which doubles as growth mould and protective shell. The pulp shell, consisting of an outer and an inner shell, can be produced efficiently with existing pulp product manufacturing technologies and define the geometry of helmet.

The mycelium shrinks slightly during the drying process. To avoid internal stresses and, in the worst case, detachment of the mycelium from the shell, the pulp shells are designed to overlap. This allows the upper shell to move with the shrinking mycelium and slide slightly into the lower shell.

Automation capability was taken into account during development. Mycelium products, which are produced in smaller quantities or more complex shapes, are filled by hand. The filling process was redeveloped to ensure that the developed pulp moulding method meets the requirements for further development towards automation. The pulp shells are filled by injection. For this purpose, the pulp shells are held in place by a mounting device and the mycelium substrate is blown into the shell with compressed air. Small air outlets are required at critical locations to ensure that the substrate is distributed evenly everywhere. After filling the pulp mould, the growth process starts.

2.4 Environmental Life Cycle Performance

The environmental performance was analyzed with a fast-track LCA of the different material concepts. The helmet's newly conceived material concepts (helmet body plus shell/coating) were compared with conventional bicycle helmets. The following variants of conventional helmets were selected for comparison:

1. Body made of expanded polystyrene (EPS), shell made of acrylonitrile butadiene styrene (APS): EPS/ABS
2. Body made of EPS, shell made of polycarbonate (PC): EPS/PC.

The quantities for the life cycle inventory (LCI) are derived from calculations for the volume of a standard helmet. The modelling of the shell manufacturing processes was based on publicly available manufacturer data and assumptions.

According to the stage of development at the time, the following mycelium material concepts were considered:

1. Body made of mycelium with a shell made of bioplastic (maize starch)
2. Body made of mycelium with a shell made of flax and R*Concept Epoxy (40% biobased)
3. Body made of mycelium with a shell made of flax and gelatine.

The LCI was compiled by collecting data from the prototypes' manufacturing processes and the literature, especially for the mycelium substrate [30]. Details of the LCI and respective data sources are not part of this paper but can be requested from the corresponding authors.

The material concept with gelatine had a much higher environmental impact than all other mould/shell options—conventional and new. Subsequently, the flax/gelatine was excluded from the chart and as an option for the shell. Furthermore, the concept comparisons showed potential savings of CO_2-eq. of approx. 48% for the mycelium/R* concept epoxy and 78% for the mycelium/bioplastic helmets compared to conventional EPS/PC helmets.

Of the remaining material concepts, including the conventional ones, the mycelium with R*Concept epoxy had the highest environmental impacts (EI) when applying the ecological scarcity method, as shown in Fig. 2. This is due to the EI of the epoxy e.g. *carcinogenic substances into air, main air pollutants and PM, and water pollutants.*

Concluding, only the new helmet variant with mycelium and a bioplastic shell showed a better overall environmental performance compared to both conventional helmet options. Furthermore, the coating/shell emerges as a significant influence on the overall environmental performance. Therefore, a plant-based material should be preferred.

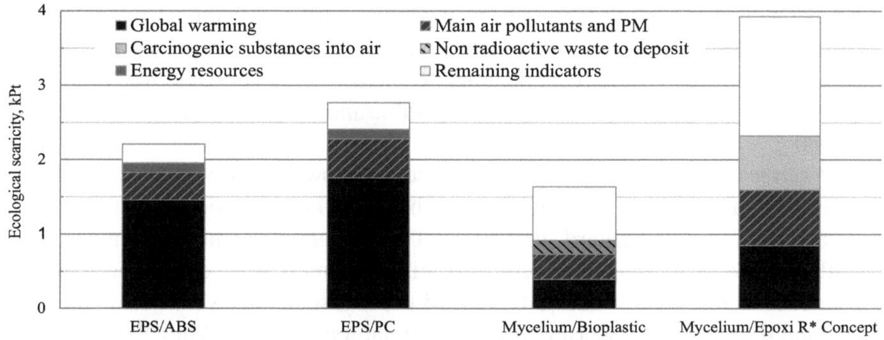

Fig. 2 Environmental impacts of helmet variants—new (Mycelium/Bioplastic, Mycelium/Epoxy R* Concept) and conventional (EPS/ABS, EPS/PC), assessed with Ecological Scarcity 2013 V1.06 in kPt at level characterization showing the main environmental impact categories per variant

2.5 *Mechanical Testing*

The final part of the proof of concept is to prove that the developed and manufactured helmet complies with the stipulated standard. For this purpose, mycelium samples and helmets in various stages of development were subjected to drop tests based on the standard [27]. Specifically, the temperature and ageing influences were not considered and the helmet was only tested for its shock absorption capacity in relation to a fall from a height of 1.479 m (5.42 m/s) onto a flat plate, based on the standard specification. To conserve the meaningfulness of the tests, the results from the manufactured mycelium samples and helmets throughout the various stages of development were compared to the results of drop tests with conventional helmets on the market. The helmets and samples were tied to a dummy test head with an internal accelerometer. The measured peak acceleration converted to multitudes of the gravitational acceleration (g) is plotted in Fig. 3. The following tests were carried out and yielded the following positive results:

- *Test series 1.* Square (140 × 140 × 40 mm) mycelium hemp samples made by hand are covered with different protective layers against external influences (e.g. UV, moisture, and dirt). The coatings were made of different blends of biodegradable materials (see Fig. 3). Samples 1.1–1.3 were coated with a pulp, gelatine and impershield blend. Samples 1.4–1.6 were coated with a pulp and impershield blend. All of the samples damped the instantaneous shock to reduce the a peak acceleration value to be below 250 g. On average, they damped the shock to 139 g.
- *Test series 2.* The results of the test series 2 is not displayed in the chart in Fig. 3. Two external companies made the first helmet prototype series. The helmets grew in the developed moulds but with the company's mycelium mixture. For the first drop test, no protective top layer was added. The polycarbonate outer layer of the two helmets ([31] and [32]) was removed for an accurate comparison. The peak acceleration of the commercial helmets was below 173 g. One mycelium

helmet had a peak acceleration of 184 g which is still within the norm and similar to helmets on the market. The other mycelium helmet was grown poorly; the delamination was visible from the outside and the hemp substrate was crumbling out. This helmet exceeded the maximum value and could only dampen the impact to 380 g.

- *Test series 3.* The 1st helmet prototype series grown from purchased substrate, consists of a mycelium hemp mixture with two variants of a protecting hardcover layer. One helmet has a hard shell made of a flax-gelatine composite and the other with a flax-epoxy composite. Both mycelium helmets were compared with two reference helmets ([33] and [34]). All of the test specimes were below the peak acceleration limit. The worn UVEX helmet damped to 211 g while the Decathlon helmet made it to 132 g (see Fig. 3, 3.1 and 3.2). The two mycelium helmets had comparable damping features and damped to 140 and 112 g.
- *Test series 4.* The final prototype (pulp shell, blown in) was tested twice. Once as a new helmet and once more as an already deformed helmet to validate the measured values. Due to an insufficient quantity of final prototype helmets, the double test method was applied to this helmet. The first impact showed an acceleration of 171 g (see Fig. 3, 4.1), the second impact exceeded the limit with a messured acceleration of 265 g. This shows that even if the helmet is already damaged, it still has high damping features the lie almost within the required norm.

The drop tests with all the well grown, undamaged mycelium specimens resulted in a peak acceleration below the limit of 250 g. Indeed, the results were even below 170 g, which provides a large safety margin that buffers possible test and manufacturing tolerances and can be incorporated into the design specifications.

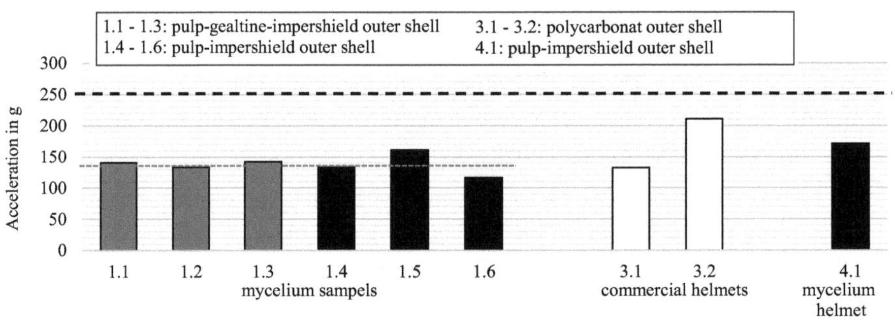

Fig. 3 Shock absorbing capacity test for fall on flat plate

3 Discussion

The study confirms the fundamental feasibility of a mycelium-based bicycle helmet and provides evidence to adapt the Design for X (DfX) processes accordingly. Mechanical, technical, procedural (production), and sustainable requirements are fulfilled. With the combination of mycelium and paper, the Cradle to Cradle principle could be achieved in case it is ensured that the paper is biodegradable and/or compostable [35, 36]. The approach to DfX has been holistic, confirming that the product development process must change significantly. Therefore, companies are challenged to withstand the adapted standards, e.g. to align with the EU's circular economy action plan [3].

To keep a common consumer product uncompromisingly in the biological cycle, there are a considerable variety of challenges and changes in the design process required. It is essential to integrate DfE and DfM in an early design phase because on the one hand the ecological effects are not yet known and on the other hand DfE significantly influences material definition, product design, production and customer acceptance. An iterative process across all DfXs is essential to find the optimum solution for all DfXs. In this case, DfE had an enormous impact on DfM and DfA. The manufacturing process was completely revised and harmonised with the materials selection, the ecological impact, economic efficiency and customer acceptance. Suitable parameters and processes had to be found and defined within DfM.

The holistic approach makes it possible to get insights into all areas, and thus develop overarching solutions. However, detailed studies and tests were not carried out for every Design for X aspect, which is necessary for the serial implementation of the processes. Therefore, the following points need to be investigated further:

- Process reliability
  - Reliable, leak-proof filling of the shells by an automated system
  - Adaption of the bicycle helmet shape and volume to compensate the shrinkage behaviour of the mycelium during the drying process
  - Reiteration of the fastening system integration and related optimization
  - Scaling and ensuring constant quality to achieve repeatability of the production results by process monitoring and control despite being faced with a bio-based, and thus changing material
  - Qualification of the raw material to adapt and stabilize production parameters by testing and consideration of the different environmental influences
  - Investigate the long-term behaviour and durability of the products (wear and tear)
- Design
  - Investigate customer acceptance through surveys to adapt the product
- Mechanical Testing
  - Testing bicycle helmets in a quantity to achieve statistical significance

- Evaluating not-considered elements
  - Optimizing the fastening system with regards to the cradle to cradle approach
  - Achieving the economic viability of the mycelium bicycle helmet (bioeconomy).

References

1. 'A European Green Deal'. Accessed: Aug. 18, 2023. [Online]. Available: https://commission.europa.eu/strategy-and-policy/priorities-2019-2024/european-green-deal_en
2. Directorate-General for Research and Innovation (European Commission) *et al.*, *Industry 5.0, a transformative vision for Europe: governing systemic transformations towards a sustainable industry*. LU: Publications Office of the European Union, 2021. Accessed: May 10, 2023. [Online]. Available: https://data.europa.eu/doi/10.2777/17322
3. 'Circular economy action plan—European Commission'. Accessed: Feb. 14, 2024. [Online]. Available: https://environment.ec.europa.eu/strategy/circular-economy-action-plan_en
4. 'Schweiz-EU in Zahlen'. Accessed: Aug. 21, 2023. [Online]. Available: https://www.eda.admin.ch/europa/de/home/dienstleistungen-publikationen/schweiz-eu-in-zahlen.html
5. 'Die Tech-Industrie im nationalen Kontext', Panorama. Accessed: Aug. 21, 2023. [Online]. Available: https://panorama.swissmem.ch/de/aktuelle-kennzahlen/die-tech-industrie-im-nationalen-kontext.html
6. 'Exportzahlen', Panorama. Accessed: Aug. 21, 2023. [Online]. Available: https://panorama.swissmem.ch/de/aktuelle-kennzahlen/exportzahlen.html
7. H. Meffert, C. Burmann, M. Kirchgeorg, and M. Eisenbeiß, *Marketing: Grundlagen marktorientierter Unternehmensführung Konzepte—Instrumente—Praxisbeispiele*. Wiesbaden: Springer Fachmedien, 2019. https://doi.org/10.1007/978-3-658-21196-7
8. W. McDonough, M. Braungart, and W. McDonough, *Cradle to cradle: remaking the way we make things*. New York: North Point Press, 2002
9. 'Circulate products and materials'. Accessed: Feb. 15, 2024. [Online]. Available: https://www.ellenmacarthurfoundation.org/circulate-products-and-materials
10. *VDI 2221 Blatt 1 & 2—Entwicklung technischer Produkte und Systeme*. 2019. Accessed: Oct. 17, 2022. [Online]. Available: https://www.vdi.de/richtlinien/details/vdi-2221-blatt-1-entwicklung-technischer-produkte-und-systeme-modell-der-produktentwicklung
11. K. Navare, B. Muys, K. C. Vrancken, and K. Van Acker, 'Circular economy monitoring—How to make it apt for biological cycles?', *Resour. Conserv. Recycl.*, vol. 170, p. 105563, Jul. 2021, https://doi.org/10.1016/j.resconrec.2021.105563
12. S. Filippi and I. Cristofolini, *The Design Guidelines Collaborative Framework*. London: Springer, 2010. doi:https://doi.org/10.1007/978-1-84882-772-1
13. K. Bauer, S. Schick, A. Wagner, K. Zhou, S. Peldschus, and A. Malczyk, *Untersuchungen zur Schutzwirkung des Fahrradhelms*, vol. 32. in Forschungsbericht, no. Unfallforschung der Versicherer, vol. 32. Berlin: Gesamtverband der Deutschen Versicherungswirtschaft e. V., 2015. [Online]. Available: https://www.rechtsmedizin.med.uni-muenchen.de/service/downloads/forschungsbericht_fahrradhelm.pdf
14. 'Parts of a bike helmet—Trek Bikes (GB)'. Accessed: Feb. 16, 2024. [Online]. Available: https://www.trekbikes.com/gb/en_GB/bike-helmet-parts/
15. K. A. G. & C. K. Pfarrkirchen D.-84347, 'Herstellung eines Fahrradhelmes'. Accessed: Feb. 12, 2024. [Online]. Available: https://fahrrad.news/journal/herstellung-eines-fahrradhelmes__10.php
16. N. M. Holden, A. M. Neill, J. C. Stout, D. O'Brien, and M. A. Morris, 'Biocircularity: a Framework to Define Sustainable, Circular Bioeconomy', *Circ. Econ. Sustain.*, vol. 3, no. 1, pp. 77–91, Mar. 2023, https://doi.org/10.1007/s43615-022-00180-y

17. E. C. D. Tan and P. Lamers, 'Circular Bioeconomy Concepts—A Perspective', *Front. Sustain.*, vol. 2, 2021, Accessed: Feb. 15, 2024. [Online]. Available: https://www.frontiersin.org/articles/10.3389/frsus.2021.701509
18. 'Circular economy: strategies and policies', *Field Actions Sci. Rep.*, vol. 23, pp. 4–7, 2021
19. S. Ura and S. Shahinur, 'Developing Successful Biobased Product: Key Design and Manufacturing Challenges', 2020
20. T. Balaeş, B.-M. Radu, and C. Tănase, 'Mycelium-Composite Materials—A Promising Alternative to Plastics?', *J. Fungi*, vol. 9, no. 2, p. 210, Feb. 2023, https://doi.org/10.3390/jof9020210
21. A. Van Wylick, E. Elsacker, L. Yap, E. Peeters, and L. De Laet, *Mycelium Composites and their Biodegradability: An Exploration on the Disintegration of Mycelium-Based Materials in Soil.* 2022. https://doi.org/10.4028/www.scientific.net/CTA.1.652
22. W. Sun, M. Tajvidi, C. Howell, and C. G. Hunt, 'Functionality of Surface Mycelium Interfaces in Wood Bonding', *ACS Appl. Mater. Interfaces*, vol. 12, no. 51, pp. 57431–57440, Dec. 2020, https://doi.org/10.1021/acsami.0c18165
23. Pré Consultants, *SimaPro.* (2021). Pré Consultants. [Online]. Available: https://pre-sustainability.com/solutions/tools/simapro/
24. G. Wernet, C. Bauer, B. Steubing, J. Reinhard, E. Moreno-Ruiz, and B. Weidema, 'The ecoinvent database version 3 (part I): overview and methodology', *Int. J. Life Cycle Assess.*, vol. 21, no. 9, pp. 1218–1230, Sep. 2016, https://doi.org/10.1007/s11367-016-1087-8
25. G. Montgomery, U. Nöthiger-Koch, and H. H. Schmitt, 'Swiss Eco-Factors 2013 according to the Ecological Scarcity Method', *Environ. Stud.*, 2013
26. T. IPCC, *Climate change 2013: the physical science basis: summary for policymakers, a report of working group I of the IPCC: technical summary, a report accepted by working group I of the IPCC but not approved in detail: and frequently asked questions: part of the working group I contribution to the fifth assessment report of the intergovernmental panel on climate change.* s.l.: WMO, UNEP, 2013
27. 'DIN EN 1078:2014-04, Helme für Radfahrer und für Benutzer von Skateboards und Rollschuhen; Deutsche Fassung EN_1078:2012+A1:2012', Beuth Verlag GmbH. doi:10.31030/2101452
28. *Mycrobez, Grellingerstrasse 27, CH-4052 Basel,* https://mycrobez.ch/
29. M. A. Elsawy, K.-H. Kim, J.-W. Park, and A. Deep, 'Hydrolytic degradation of polylactic acid (PLA) and its composites', *Renew. Sustain. Energy Rev.*, vol. 79, pp. 1346–1352, Nov. 2017, https://doi.org/10.1016/j.rser.2017.05.143
30. F. J. Leiva, J. C. Saenz-Díez, E. Martínez, E. Jiménez, and J. Blanco, 'Environmental impact of Agaricus bisporus mycelium production', *Agric. Syst.*, vol. 138, pp. 38–45, Sep. 2015, https://doi.org/10.1016/j.agsy.2015.05.003
31. *Rockrider MTB ST 100, seller: Decathlon*
32. *BTWIN City 100, seller: Decathlon*
33. *BTWIN Bowl 500, seller: Decathlon*
34. *UVEX Bike Helmet*
35. Y. Zhang, C. Duan, S. K. Bokka, Z. He, and Y. Ni, 'Molded fiber and pulp products as green and sustainable alternatives to plastics: A mini review', *J. Bioresour. Bioprod.*, vol. 7, no. 1, pp. 14–25, Feb. 2022, https://doi.org/10.1016/j.jobab.2021.10.003
36. 'Is Paper Biodegradable: Nature's Recyclable Material—Eco Sustainity'. Accessed: Sep. 03, 2024. [Online]. Available: https://ecosustainity.com/is-paper-biodegradable/

From Reviews of Sustainable Building Rating Systems and Circular Indicators for the Built Environment to a Practical Tool Supporting the Design of Sustainable Circular Buildings

Leidy Guante Henriquez, Francesco Frontini, Marco Imperadori, Carlo Gambato, and Stefano Zerbi

Abstract The Construction sector is one of the main contributors to climate change and resource scarcity we face today. The transition towards Circular Economy (CE) models, and its interpretation at the building level with the Circular Building (CB) idea, represents a possible solution. Unfortunately, previous research has reported the lack of a uniform and well-established performance assessment scheme to define and measure CBs. Considering the potential for impact during design decision-making, this study estimates opportunities for integrating circularity principles into the building design process and, on that basis, defines the groundwork for the future development of a meta-protocol supporting the planning of efficient and sustainable CBs. The paper presents the underlying framework and base structure of the in-progress tool by reviewing existing Sustainable Building Rating Systems' (SBRSs) integration of CE principles and analyzing developed metrics for practical circularity assessment in the Construction sector.

Keywords Circular building · Design tools · Circularity · Sustainability · Assessment methods · Circular economy (CE) · Sustainable building rating systems · Sustainable building certifications

L. Guante Henriquez (✉) · F. Frontini · C. Gambato · S. Zerbi
SUPSI, DACD, Mendrisio, Switzerland
e-mail: leidy.guantehenriquez@supsi.ch

L. Guante Henriquez · M. Imperadori
ABC Department, Politecnico Di Milano, Milano, Italy

A. Podleisek et al. (eds.), *Advances in Circular Economy: Bridging Research and Practice*, Springer Proceedings in Earth and Environmental Sciences,
https://doi.org/10.1007/978-3-032-21547-5_7

1 Introduction

1.1 Current Scenario

In Europe, buildings are responsible for 40% of energy consumption, 36% of greenhouse gas emissions, 50% of extracted material resources, and 35% of total waste generation [1, 2]. Furthermore, the "Roadmap to a Resource Efficient Europe" [3] pointed Construction Industry as a crucial sector, highlighting that better design and use of buildings can contribute to resource saving. And, according to the "Circular Economy Action Plan" (COM/2020/98), a way to achieve that is the transition of the sector to a Circular Economy (CE) [4].

In contrast to the traditional linear economy, CE is restorative and resilient and aims to eliminate or strongly reduce waste, circulate materials, and regenerate nature [5]. At the building level, the following approach can be translated into the idea of a "Circular Building" (CB), defined as a "building that is designed, developed, used, and reused without unnecessary resource depletion, environmental pollution, or ecosystem degradation. It is constructed in an economically responsible way and contributes to the well-being of people and the biosphere. {…}Technical elements are demountable and reusable, and biological elements can also be brought back into the biological cycle" [6].

1.2 Problem Statement

At present, there are several barriers to the CB model. Specifically, previous researchers have reported the lack of a uniform and well-established performance assessment framework to define and measure CE in the built environment [7], whilst policymakers and international standards have emphasized the need for a universally recognized circularity framework for buildings [8]. Beyond that, the literature highlights significant obstacles as designers' and clients' concerns with the CE approaches [9, 10].

Furthermore, considering that the ability to influence the functional capabilities of a project while maintaining low costs is higher during the first process phases [11], stakeholders involved during design usually represent drivers boosting bottom-up innovation in the building sector. Future studies should focus on the development of improved replicable models for the implementation of CE during building design and the determination of standardized indicators to qualify CBs.

Secondly, many authors investigated the controversy concerning the link of CE to Sustainability [12]. The dominant framing places CE on one line with other sustainable strategies, making circularity a concept complementing conventional principles rather than being the first sustainability concept [12] and placing Sustainable Development as the primary goal. Nevertheless, the CE concept is still determinant in developing a new consumption culture [13] and, in that sense, should be considered a valuable instrument for Sustainable Development with a certain margin of ontological autonomy, but compliance is not so obvious [14]. Furthermore, research should focus more on the demonstration of actual sustainability of all the circular strategies [15], going beyond carbon emissions considerations. In this respect, instead of venturing into completely new evaluation schemes, it should be considered that the Sustainable Building archetype is a fully developed and approved concept framed by available Sustainable Building Rating Systems (SBRSs). In fact, a SBRS can influence many choices during a building lifecycle [15] and even currently features some hints of CE concepts thus, becoming useful tools for fostering and implementing circular approaches in the building sector.

Thirdly, Academia and professionals developed metrics (circular indicators) to quantify circularity integration into the Built Environment. These elements represent significant instruments for CB assessment because incorporate CE ideas of resource minimization and cycling in the product design and management while providing high synthesis and communication of CE core features [16]. However, still uncertain how to connect them to Sustainability or use them in comprehensive evaluations of CBs.

In conclusion, in the transition toward CE, the building sector needs accessible working tools to make logical and straightforward for stakeholders involved during the design stage the idea of "thinking circularly". Sustainability can be used as a reference paradigm to confirm the advantages of circularity, and SBRSs can be a starting point for developing shared criteria to measure sustainable CE assimilation in constructions, while circular indicators can be used to operationalize and simplify circularity measurements. Accordingly, the paper provides the theoretical basis to elaborate an efficient and ready-to-use tool able to support the decision-making involved during the design of building projects implementing sustainable-focused CE concepts.

1.3 State of the Art: Circularity Assessment for the Built Environment

As a first step, we reviewed current academic advancements in circularity assessment for the Construction sector. Initial approaches to CE adoption measures, policies, and frameworks were at a "macro-level" (i.e., urban agglomerates). Buildings can be seen as a "micro-level" and building components and materials are the "nano-level" [8]. Given the purpose of the study, we will focus on the building level. Several building-specific indicators are available, but they lack maturity and are often not validated. During 2013–2018, researchers addressed the challenges of CE-enabled design and tried to address its implications on Sustainability. Instead, since 2016, publications dealing with potential methodologies for CE evaluation [17] have been released. At that point emerged the need to develop measurement tools for understanding the circularity degree of buildings. Circularity assessment as a trending topic in the last years led to a proliferation of metrics referring to different paradigms for the evaluation of CBs and their components. The phenomenon, in the long run, can create confusion, failing the purpose of helping designers and planners comprehend circular strategies. Nevertheless, circularity assessment appears indispensable to fostering CE-related improvement actions and effectively communicating them to others [18].

Specifically, to assess CE strategies, several building-specific indicators are available, but they are incomplete and immature, being not enough tested. A detailed bibliographical search highlighted the lack of proper discussions on the correlation between circularity and sustainability measurements, especially when looking at CE implications on the social dimension of Sustainability [14, 19, 20]. So far, Academia has tried to address the issue by developing new assessment frameworks or metrics coupling CE and Sustainable Building principles.

Some researchers attempted to comprehensively introduce the sustainability concept to the CBs assessment debate. At present, no common or harmonized circularity assessment framework has been found, and proper multi-criteria evaluation approaches combining CE and sustainability considerations to improve the building life cycle decision-making still need to be included. Dams et al. [21] elaborated a Circular Construction evaluation framework that assesses structures' adaptability and flexibility through sets of criteria in a tabulated format (CCEF) and tries to consider some social aspects ("Health and safety" requirements) while encouraging Life Cycle Assessment (LCA) considerations. Gonzàlez et al. [22] presented a methodology to assess the degree of circularity of construction/renovation works in a life cycle perspective, calculating five indicators: "Energy circularity", "Materials circularity", "Water circularity", "Social added value" and "Life cycle economic value". The methodology properly represents the multi-disciplinary feature of CE; however, sustainability principles are used only to assess the same circularity concept through the five mentioned parameters. Instead, a clear distinction between circularity and sustainability metrics is necessary for formulating a comprehensive CE assessment: some indicators should measure the intrinsic circularity degree of a system, and others evaluate the sustainability of CE strategies and the overall project [23]. Put differently, in the proposed method, the link between the two paradigms remains implicit, preventing comparison, which is essential to improve the understanding of circular solutions. Larsen et al. [24] stated that for CE to succeed, a comprehensive and circular view of buildings' life cycle phases is necessary. Therefore, they proposed a Life Cycle Sustainability Assessment (LCSA), an evaluation integrating LCA, Life Cycle Costing (LCC), and Social Life Cycle Assessment (S-LCA). The study also highlights the relevance of the social dimension of Sustainability, which is usually neglected when measuring circularity. However, considering the complexity of life cycle methodologies and the little application of S-LCA in the Construction sector, the proposed approach needs further maturation and development. Instead, Hasheminasab et al. [25] developed a comprehensive assessment framework for construction projects linking all dimensions of Sustainability (economic, environmental, and social) with CB principles by exploiting existing SBRSs. Unfortunately, the resulting assessment framework seems intricate. Firstly, recognized multi-criteria decision-making methodologies (AHP[1] or ITARA[2] and PMADM[3]) are combined to define on a case-by-case basis weighting criteria according to the specific condition of the analyzed project (e.g., building typology). Then the best building project solution is selected through the CoCoSo[4] method and a decision matrix based on the experts' opinions. Rather than exploiting certifications and other sustainable building rating systems to prove the win–win feature of circular strategies for sustainable buildings, the proposed assessment searches for the best solution in terms of investments. The solutions of the last two references represent a good direction for the combination of CE and sustainability evaluations; however, they are also quite complex, time-consuming, or not usable in practice. When focusing on decision-making in the design stage, a more structured and intuitive method is needed to assist involved stakeholders properly. New variables and issues related to circularity involve a higher degree of complexity that should be counterbalanced by a targeted set of procedures facilitating building design while limiting efforts. The solution should be readily usable, not overly time-consuming, and easily integrated into the current design process.

The next section presents the broader research outcome and reviews existing frameworks/metrics to set up the theoretical background for a so-defined solution.

2 The Outlined Research Outcome: A Meta-Protocol for Sustainable Circular Building Design

This study is part of a Ph.D. research project aiming to estimate opportunities for integrating circularity principles into the building design process and, on that base, define a meta-protocol for planning efficient and sustainable CBs.

Taking account of current metrics deficiencies in integrating Sustainability, better arrangements could arise by including circular indicators within sustainability assessment frameworks already in use [26]. Specifically, a possible solution could come from a synthesis process on the building circularity indicators developed until now and criteria from SBRSs. The idea is to take advantage of already accepted schemes to assess sustainable buildings and avoid the proliferation of further procedures that still need complete validation. Furthermore, the proposed scheme will support the design of sustainable CBs during each decisive step of the design process (Fig. 1).

[1] "Analytical Hierarchy Process" is a method that supports multi-criteria decision-making confronting more alternatives in relation to a plurality of criteria, quantitative or qualitative.

[2] "Indifference Threshold-based Attribute Ratio Analysis" is another method that supports multi-criteria decision-making that can be used to generate reliable weights for risk factors.

[3] "Prospective Multiple Attribute Decision Making" is a novel approach to cope with future uncertainties during multi-criteria decision-making.

[4] "Combined Compromise Solution" is a method supporting multi-criteria decision-making by computing different weighting matrices and calculating various relative weight measures.

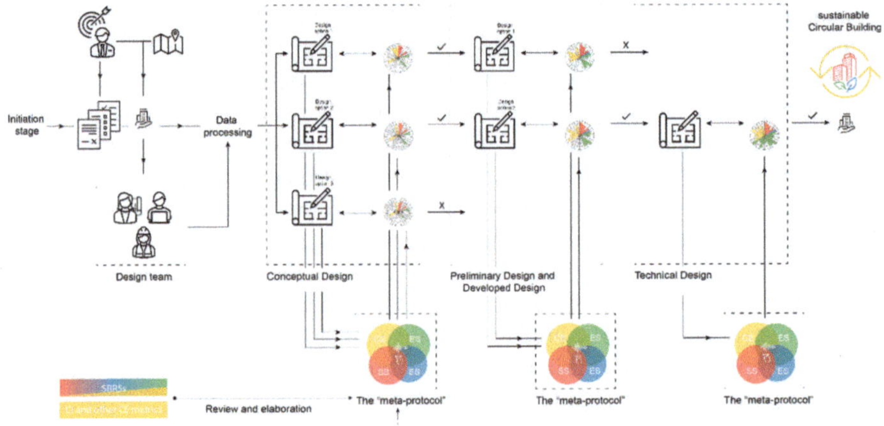

Fig. 1 Schematization of a building design process involving the foreseen meta-protocol

Considering data conventionally available at each "sub-stage",[5] reviewing available metrics for circularity and sustainability evaluation, and collecting opinions from Construction Industry experts and practitioners, we will define a set of criteria, indications, and methods that should be followed for creating structures that integrate CE without compromising sustainability concerns. The project will improve the design process of CBs by providing immediate feedback on the success of implemented CE strategies. Furthermore, the meta-protocol base framework could represent a baseline for sustainable building certifications in complementing their rating systems (with additional requirements related to CE) or can be applied by policymakers in the development of tools/guidelines for CBs. Finally, the project could significantly impact the Construction sector, changing some main stakeholders' (e.g designers and clients) perceptions of CE and promoting related solutions for the Built Environment through the demonstration of circularity compliance with sustainability requirements defined by nowadays policies and official accreditations.

2.1 Research Methodology

This section illustrates the methods used to attain the research goal; the approach is inspired by the Design Research Methodology (DRM [27]). Accordingly, the research project is structured in four main stages: the Research Clarification, the Descriptive Study I, the Prescriptive Study, and the Descriptive Study II. In the Research Clarification stage, we identify evidence in support of current metrics and

[5] CEN, Engineering services—Terminology to describe engineering services for buildings, infrastructure and industrial facilities (EN 16310:2013), (2013).

frameworks' inconsistency for assessing CBs (main findings summarized in paragraph 1.3), especially related to Sustainability. In the first Descriptive Study, with detailed literature reviews, we define the meta-protocol framework. In the third phase, we will complete the meta-protocol protocol and associated tools to facilitate its application. In the second Descriptive Study, the novel meta-protocol will be then applied to more than two case studies to verify its applicability and usefulness in supporting CE implementation during building design and thus in fostering CB model.

This paper describes efforts to achieve valuable findings from the first Descriptive Study. Specifically, the study reports on the detailed literature review and the associated critical work results needed to outline the architecture of the meta-protocol.

2.2 *Existing Building Sustainability Assessment Frameworks*

The development of systems able to evaluate the compliance of buildings with the Sustainability paradigm represents a key point for the transition of the Construction Industry towards a more conscious and environmentally friendly configuration. By formalizing the design and performance criteria of sustainable buildings, Green Building Rating Systems can boost innovation and normalize good practices toward sustainability [28]. Nowadays, around 600 Green certification systems for products are available [29] and at least 50 apply to buildings. Initially, they focused more on the environmental aspect. Still, lately, social criteria have shown a significant growth trend, and the economic dimension is more addressed, highlighting the willingness of a more balanced assessment. That is why it is now possible to call some of these schemes Sustainable Building Rating Systems (SBRSs) or Sustainable Building Certifications [30].

Therefore, relevant SBRSs in the European Construction sector and some rating systems recognized at the international level were examined. Subsequently, clear, accessible, transparent assessment frameworks, with the ability to investigate more than one aspect of sustainability [28], were selected for a critical review concerning the type of aspects investigated, related dimensions of sustainability, circularity principles incorporation, and where indicators associated with selected criteria involve quantitative or qualitative evaluations (Fig. 2). Besides the conventional dimensions of sustainability (social, economic, and environmental), SBRS's criteria were classified according to the CE category. If related indicators measure or describe strategies related to material efficiency (narrowing loops), reuse, remanufacturing, or extending product durability (slowing loops), plus recycling, energy recovering, and waste management (closing loops), they introduce circularity topics. The following sub-paragraph illustrates how the selected SBRSs integrate circularity concerns, highlighting strengths and gaps, while Fig. 2 summarizes the analysis results.

Active House. Active House is a recent Danish certification system developed in 2017 [31]. It relies on three main principles: Comfort, Energy, and Environment,

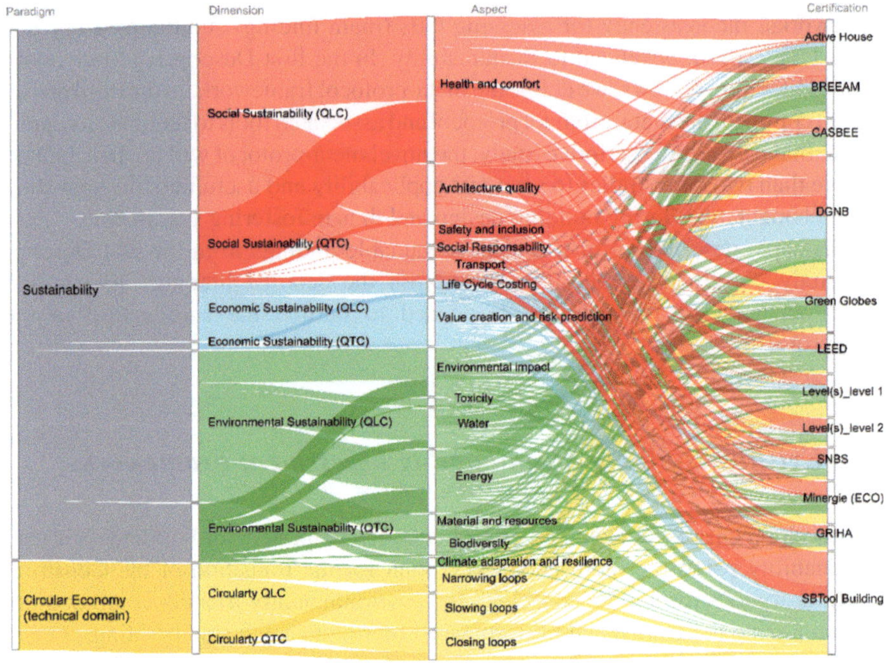

Fig. 2 Flow scheme showing the proportion of aspects and principles investigated by the analyzed sustainable building rating systems (SBRSs)

so it mainly focuses on environmental and social sustainability. A few CE-related criteria are included in the last version of the scheme: one about building disassembly and two related to minimum values of average recycled/recyclable or reused/reusable content for all building materials [32]. Active House assessment framework is intuitive thanks to a limited number of straight criteria and associated indicators, the unambiguous distinction between required quantitative and qualitative evaluations, and the simplicity of the scoring system. Another advantage of the Active House is the efficient summarization of the obtained results through a "radar". This readily intelligible tool effectively communicates, understands, and manages the design and operational choices. That makes the Active House framework an ideal base to measure and monitor parameters related to sustainability but also circularity. The great potential of the AH radar as an pioneering design and monitoring tool is dictated by the fact that the Active House certification, although essential, is well structured and gives the same weight to all the criteria considered. The study conducted by Di Santo et al. [33] has indeed used this certification scheme together with LCA methodology and BIM to improve the design of sustainable buildings. The limit of Active House is that economic sustainability should be more addressed.

BREEAM. The Building Research Establishment Environmental Assessment Method (BREEAM) was the first developed Sustainable building certification system. Created in 1990 by the Building Research Establishment (BRE) group, BREEAM is currently one of the most used certification protocols in the world [34]. Its assessment framework is characterized by accurate indicators investigating mainly sustainability's environmental and social aspects through ten thematic categories. "Materials" and "Waste" categories are the most relevant in CE discussion. The first category introduces circularity concerns with "Material Efficiency" and "Designing for durability and resilience" criteria. The latter encourages the adequate protection of building elements to minimize the frequency of repairs. Instead, many criteria in the "Waste" category foster the use of recycled aggregates, promote efficient waste management during construction/operational phases, and discourage premature demolitions through adaptability principles [35]. Nevertheless, BREEAM certification still presents significant limits when assessing the full implementation of CE in a building. The report "A Framework for Circular Buildings" precisely concerns the inclusion of an assessment framework for CBs into BREEAM and suggests solid strategies and indicators for their possible inclusion in "New Construction" and "Refurbishment and Fit-out" schemes [6]. The document, starting with an interesting CB definition, proposes a strict correlation between CE and Sustainable Development, not only emphasizing the avoidance of unnecessary waste of resources but also considering the protection and growth of natural capital (biodiversity), social capital (well-being, occupant safety, society, and culture humanity) and the creation of other forms of value. For example, material toxicity is significant when considering reuse because this feature is a prerequisite for reintroducing a product in a technical cycle while ensuring a healthy environment. Regarding durability issues, another erroneously overlooked aspect is the aesthetic value of a building, which becomes important as it contributes to the overall value of society. The authors also point out CB aspects that are not yet covered by the standard. Some fundamental strategies, such as Design for Disassembly, are missing, or some considerations, such as the availability of materials information along the life cycle of the building, need to be improved.

CASBEE. The Japanese certification "Comprehensive Assessment System for Built Environment Efficiency" (CASBEE) [36] also incorporates some CE concerns. Specifically, it inspires slowing loop approaches by demanding high level of maintenance and pushing on the evaluation and, consequently, on the elongation of repair and replacement intervals. It also reports different criteria concerning building durability that include assessing reliability during extreme natural events (e.g., earthquakes). Furthermore, CASBEE provides clear practical solutions to allow functional adaptability and flexibility of spaces, promote the use of recycled materials, empower "Adaptive Reuse", and qualitatively measure efforts to enhance the reusability of building components suggesting Design for Disassembly (DfD) concepts [37].

DGNB. The "Deutsche Gesellschaft für Nachhaltiges Bauen" is a system for the certification of buildings created by the German Sustainability Council in 2007 [38]. Among all the above mentioned, DNGB is the certification that allows the more holistic analysis of all sustainability dimensions [28, 39] as it pays more attention to economics. In 2019 DGNB drew up the report "Circular Economy—Closing Loops

Means Being Fit for the Future" [40]. The report underlines the fundamental role of the CE in reducing consumed resources, produced emissions, and waste characterizing the Construction sector. They provide a toolbox to better implement CE strategies in the Built Environment and estimate the current state of circularity consideration in the standard. In DGNB, CE principles are already foreseen in different criteria and associated "bonus points", plus added credits for CE experimentations. Bonus points concerning "Life cycle cost" are assigned if it can be proven that a relevant proportion of building components have been reused in the project or used by exploiting circular business models. The criterion concerning "Value stability" involves additional points if users, tenants, or hires of the building actively contribute to the Circular Economy. "Documentation" incentives the development of data basis for future Urban Mining activities (e.g., a "Building Resource Passport"), BIM2FM[6] concepts implementation, and digitization of existing structures. "Use and integration of building technology" rewards energy-sharing models at the district/local level, while "Ensuring sustainability aspects in tendering and contracting" gives bonus points if the invitation to tender contemplates mineral-based recycled materials or explicitly requests reused/recycled material to manufacture mineral-based construction products. Finally, the last version of DGNB system includes a completely circular-oriented criterion: Circular Construction (TEC1.6). The criterion promotes the use and the conservation of existing buildings, distinguishing whether previous deconstruction can be assigned to the project and pushing for circular principles integration from early design stages. It also rewards circularity documentation and implementation by assessing key metrics ("Primary raw materials avoided", "(Circular) material origin—Closed-loop recycling implemented", "Reduction and recycling of construction and demolition waste", "Circularity—post-use pathways"). Furthermore, if reclaimed elements are used in the site, or if quantities assessment of the overall building is performed and then reported in the complete "Building Resource Passport", bonus points can be credited.

Green Globes. Canadian standard Green Globes is a science-based rating system that evaluates the environmental sustainability, health, wellness, and resilience of commercial buildings [41]. Concerning CE topics, the certification includes different waste-focused criteria ("Construction waste", "Post Occupancy Solid Waste Recycling", and "Supply Chain Waste Minimization" criteria) and materials use minimization through modularity concepts and/or prefabrication ("Off-Site Fabrication for Construction Optimization"). Additionally, Green Globes with the "Design for Deconstruction" criterion recommends documenting the application of design for deconstruction (DFD) principles in the building design and providing the client with the related Deconstruction Plan. It also suggested the reuse of existing building components providing points for a certain percentage of structural elements and non-structural interior systems retained in a new design [42].

---

[6] "BIM2FM", i.e., Building Information Modelling to Facility Management; the term is usually exploited to indicate the smooth transition of BIM data from the planning and construction phase to the operational phase of a built asset.

GRIHA. The Green Rating for Integrated Habitat Assessment was born as an Indian standard for Green Buildings [43]. The last version attempts to qualify environmental concerns, user experience, and market feedback, integrating life cycle cost analysis, life cycle analysis, and water performance index concepts. From a CE perspective, GRIHA wants to minimize a building's resource consumption, waste generation, and overall ecological impact, mainly considering nationally acceptable limits/benchmarks. The most significant points are present in the "Construction Management Practices" criterion (n.6) encouraging CDW minimization, the "Solid Waste Management" section's criteria (n.17 and 18) mostly concerning post-occupancy waste, and the "Sustainable Building Materials" ones promoting the use of alternative materials able to conserve natural resources (n.19 and 21) [44].

LEED. LEED (1998) is one of the most widespread existing certification systems. In LEED v4.1, most of the credits are related to operational and embodied carbon, with about 35% of them related to climate change, 20% to human health, 15% to water resources, 10% to biodiversity, 10% to green economy concepts, and a smaller percentage to impact on community and natural resources [45]. Surprisingly, LEED has one of the least circular-oriented rating systems (Fig. 2). Few principles begin to be introduced through criteria related to the minimization, selection, accountability, and efficient use of building products throughout the project life cycle, design for flexibility, plus construction and demolition waste management planning [46].

Minergie (ECO). Minergie is a Swiss certification for sustainable construction that since the 1990s has certified over 55,000 buildings. It aims to promote energy-conscious construction that pursues optimal living comfort. Minergie consists of three building standards (Minergie, Minergie-P, Minergie-A) that, with different degrees of completeness, guarantee the long-term value of the building focusing mainly on energy conservation, renewable energy use, and comfort requirements. The Minergie ECO complement, in combination with one of the three Minergie standards, leads to a particularly healthy building, compatible with the circular and ecological design by introducing flexible construction concepts, careful choices of materials, and attention to water and biodiversity issues [47]. Specifically, in the 2023.1 version of Minergie ECO, CE concerns are incorporated through 13 criteria in the Ecology section. The most original ones are: (i) 210.03 which requires the assessment of the "Ecological residual value of existing structures to be dismantled" in new construction projects; (ii) 220.04 that, based on the extent of the dismantling in a construction project, asks an exanimation of the construction elements and building materials to be dismantled and/or used to verify their recyclable and reusable potential and their effective re-circulation; (iii) 220.05, 220.07, and 220.08 which, going into great detail, penalize the use of waterproofing or filling of cavities using foams, of coverings in sheets or synthetic resin mortar, of liquid synthetic sealants and of organic/mineral composite materials for which there is no recycling infrastructure; (iv) 220.09, 220.10 and 220.11 which encourage the systematic use of recycled concrete. Finally, a relevant issue can be spotted in the 210.08 criterion that highlights the importance of using local materials when implementing ecological-oriented choices. The availability of local resources is not strictly related to CE; however, during our current research activities on the topic, it has proven to be a significant element. The contextualization of circular choices should not be overlooked [48].

SBTool (Building). The SBTool is a framework for rating the sustainable performance of buildings and projects developed by iiSBE, the International Initiative for a Sustainable Built Environment. One of the most innovative features of this framework is the flexibility of the involved evaluations: (i) the system can be modified to be as narrow or as broad as desired, ranging from more than one hundred criteria to twelve; (ii) it can consider region-specific and site-specific context factors that differently influence the weight of the criteria. The SBTool's last version (2022) addresses CE concerns in around 24 criteria related to efficient use of materials, waste minimization, flexibility, adaptability, and maintenance [49].

SNBS. The Swiss certification system for buildings Standard Nachhaltiges Bauen Schweiz (SNBS) presents a well-balanced distribution of indicators referring to the three pillars of sustainability through 35 criteria [50]. The 2023.1 version delves into some CE topics with criteria 213 "Disassembly and reuse", 222 "Occupant density" (that hints at "Reduce" concepts), and 223 223 "Flexibility of use and versatility" from the Economy area [51]. When assessing a CB, the financial aspect should be addressed. Based on what has been said for Minergie (ECO), criterion 231 "Creation of regional added value" is also worthy of mention. The certification framework seems to highlight the economic benefits associated with these types of circular solutions. When assessing a CB, the financial aspect should be addressed.

Level(s). Finally, Level(s) [52] is a framework developed by the EU Commission that provides a common approach for assessing the performance of sustainable buildings by combining existing standards. This framework can be considered a reporting tool, but its results may also be analyzed to support design decisions. Level(s) consists of 6 macro-objectives and 16 performance indicators. Macro-objective "Resource efficient and circular material life cycles" directly refers to CE technical solutions. Circularity-related indicators represent a significant part of the assessment framework. CE-oriented in Level(s) framework are also the criteria defined by Macro-objective 3 ("Efficient use of water resources") and the one called "Adaptation and resilience to climate change" that evaluate building performance against projected future changes in the climate, not only to protect occupier comfort but also long-term property values [53]. Level(s) holistic approach may contribute to the CE transition of the building sector, providing a common language that supports professionals in improving building performances and policymakers or certifiers in aligning rules with environmental objectives. However, by not providing enough quantitative criteria with defined benchmarks and associated scoring systems, it can partially fail to support designers in circular decision-making, especially in comparison with sustainability concerns. The added value of this scheme is the flexibility of the level of detail at which sustainability/circularity aspects can be addressed during the different phases of a building process.

2.3 Micro Indicators and Tools for Assessing Circularity Level of Buildings

Measuring and reporting tools demonstrating advancements of CE are needed to ensure its successful transition [54] and that is why, in recent years, we have witnessed a proliferation of circularity metrics. Khadim et al. [8] carried out a detailed review of indicators formulated by Academia and enterprises to describe CE at the building (micro) and construction technologies (nano) levels. The analysis identified 18 circularity metrics to qualify the whole building. Building Circularity Indicator (BCI [55] and its updated version [56]), MADASTER[7] Circularity Indicator and FLEX 4.0 [57] are more popular ones. Most indicators are quantitative but, in their developing stage, not providing a defined or uniform way to measure circularity; specifically, a series of constraints were detected. Firstly, some of them are applicable just for a stage of analysis (e.g., SEEI [58] and IEPC [59] are just for Existing Buildings), a certain building type (e.g., Heritage Buildings [60, 61]) or a defined scale (e.g., CACE for building envelope systems developed by Finch et al. [62]). Secondly, some of them are mainly focusing on "Reuse", "Recycling", "Adaptability", and "Design to Reuse", neglecting strategies such as "Reduce" and "Repair". Abadi et al. [63] addressed the issue through a Lifecycle Circularity Assessment Framework (LCAF) that evaluates circularity during the total project life cycle and considered "Innovative Construction Methods" (e.g., Additive Manufacturing, 3D printing, off-site prefabrication …) as indicators of "Reduce" strategy during the Construction stage, plus "Durability of the Building, Assets, or Projects" as a potential indicator during the Operation stage. Thirdly, Khadim et al. [8] proceeded to state that the technical cycle is a relevant part of all the indicators, but less than half of them also include considerations about the biological loop of materials. Going further into the topic, Navare et al. [64] critically review CE assessing tools, verifying the extent to which information about biological cycles is included. Biotic materials are not necessarily circular and sustainable, and current indicators and frameworks do not properly recognize these implications. Additionally, researchers have developed assessment methods to measure energy performance and the circularity of a building in an integrated manner [59, 65, 66]. These first reflections on the environmental impact of CBs, not involving LCA, are simplified. Data on a specific supply chain, other impact indicators, and certain building life cycle stages are neglected. Furthermore, the social dimension of sustainability is generally ignored by Circular Indicators [8]. Finally, some indicators combine circularity assessment with IT-enabled computer-aided technologies like BIM. In this respect, Heisel and Rau-Oberhuber [67] evaluated the capabilities of MADASTER Materials Passports[8] (MP) to document material stocks and flows within a circular built environment and the potential of MADASTER Circularity Indicator as a design tool supporting the transition towards a circular Construction

[7] MADASTER is a platform for the registration, organization, documentation, storage, and exchange of data regarding any materials, components, and products used by the construction industry.

[8] Materials Passports are documents reporting defined characteristics of materials in products that provide them value for recovery and reuse.

Industry. Authors recognized MP as essential in providing necessary data to manage material stocks and flow within the system and of the circular indicator to summarize the performance of individual buildings in terms of resource efficiency and potential reusability.

More complex evaluation schemes were also developed. The Ellen MacArthur Foundation, in collaboration with Arup, launched the "Circular Buildings Toolkit" [68], i.e., a framework translating CE principles into a prioritized set of strategies (with associated Key Performance Indicators) relevant to planning built assets. The open and free toolkit can support designers in the integration of circular solutions in building design, also providing practical examples of the latest circular projects around the world. Instead, WBCSD[9] is conducting a project to develop a unified framework for assessing CBs. In the publication "Measuring Circular Buildings—Key Considerations" [69], they underlined some points that should be considered when evaluating circularity in the built environment. Firstly, BIM and similar technologies are essential to automate data collection and tracking materials plus related properties. Secondly, a building is a system composed of different layers, each one made of products (windows, doors, cladding, etc.) with different materials; therefore, they suggested the Element, Layer, Product, Materials (ELPMs) terminology [70] for organizing circularity-specific data. Thirdly, CE should be implemented across the building's different life phases, so a complete assessment should involve an entire building life cycle. Finally, qualitative, and quantitative data are equally important in circularity evaluations.

All these schemes are valuable references for the development of CE-specific criteria for the Built environment but for the scope of this paper more interesting are friendly-user features of developed "building circularity indicators".

To conclude, already developed micro circularity indicators for the built environment are not perfect because they mainly focus on limited aspects or are case-specific. Nevertheless, these metrics' main function is to simply represent the decoupling of economic growth from waste generation and resource depletion [16]. Thus, circularity indicators are easier to communicate and understand compared to a complete circularity assessment framework and, in combination with an exhaustive sustainability assessment framework, could efficiently support the design of CBs. The Building Circularity Indicator (BCI) or similar metrics [55, 56, 65] may be used as quantitative indicators combined with other measurements.

2.4 The Meta-Protocol's Underlying Framework

The previously mentioned analysis led to the development of the proposed supporting tool's underlying framework and base structure. The tool, a meta-protocol, will be designed to assist stakeholders involved during the decision-making of the building

[9] The World Business Council for Sustainable Development (WBCSD) is a CEO-led organization of over 200 international companies working on several issues related to sustainable development.

design process in selecting the most suitable option of CB through a series of "sub-stage-specific" evaluations.

The meta-protocol will also have a second scope: to help designers obtain an existing SBRS. Specifically, by explicitly referring to indicators, benchmarks, and the base logic of an initially selected SBRS, the meta-protocol allows designing CBs aligned with its principles thus demonstrating the consistency of Circular design with Sustainable Development. To test and demonstrate the proposed approach, Active House certification will be used because, as mentioned before, Active House is a balanced certification protocol, and its framework is essential, transparent, and intuitive. Any criteria (and associated indicators/evaluations) not available in the selected certification system but identified as a priority for CB design derive from an accurate work of synthesis on the previously analyzed SBRSs, plus other representative assessment frameworks. For instance, Active House does not provide criteria for Economic Sustainability measurements; so, that section's evaluated methods, levels, and scores are defined by looking at the ones already available in other certifications and summing up the findings according to the final goal of assessing circularity.

Researchers [9, 71] have identified domains/aspects influenced by the transition to CBs: three of them are the pillars of Sustainability (economy, environment, and society), while the rest concern logistics, role of governments and stakeholders, legal challenges, technical issues, and final users' perspective. The definition of CB domains is essential for elaborating the comprehensive circularity assessment framework underlying the proposed meta-protocol. Currently, most frameworks to evaluate CBs mainly focus on technical, environmental, and economic aspects and neglect others [72]. Instead, the envisaged assessment framework will measure and monitor the level of compliance in social, environmental, economic, and technical circularity domains while providing advantages in organizational aspects.

From findings on previous analysis and considerations, Fig. 3 reports the criteria composing the proposed meta-protocol for assessing sustainable-oriented CBs, according to the four selected domains. Based on the potentially available data, the scheme also reports at which Design phase the meta-protocol introduces the criteria. Three innovative criteria introduce novel or overlooked discourses in circularity assessment. Firstly, the "Circularity Feasibility and Local Relevance" criterion, with a context-oriented approach, deals with the logistic aspect involved in materials recirculation (e.g., the presence of recycling infrastructures, reclaimed materials exchange platforms, available demolition sites…), technological and market conditions determining CE implementation and eventual benefits to the local community associated with the project realization. The criterion aims to contextualize, during the building design, the feasibility of the circular strategies and probe the possible immediate local impacts. In CE discourses the context has a significant role. Local governments are usually formal institutions responsible for waste management. Furthermore, the use of locally available building materials is an essential strategy in the design of low environmental impact buildings [73]. That becomes even more relevant when considering waste as a resource. It allows reducing transport costs and impacts, boosts sustainable circular business models, and supports regional development. Previous research has already introduced the importance of the contexts in the assessment framework regarding buildings integrating circular strategies. For example, Parece et al. [74] developed a method for upcycled and reused building materials selection that considers the availability at the local level. Another significant point is the criterion addresses the social impacts of a CB in terms of relationships that can arise from a CE system and implications on the local community wealth (e.g., creation of new jobs, valorization of local craftsmanship, donation of surplus salvaged products, collaboration with social associations ...). Secondly, the "Supply Chain Coordination and Optimization" stresses methods to improve communication and networking of all possible stakeholders involved during a circular project. It addresses the data exchange to guarantee circular management during construction, use, and EoL of buildings and fosters the use of structured BIM procedures, digital MP, and tracking technologies allowing that through different use cycles. In fact, Giorgi et al. [15] highlighted those digital technologies are essential to trigger sustainable circular processes; however, the coupling between digitization and sustainability assessment is not enough promoted. This criterion also includes considerations that can emerge, during design, due to the adoption of circular business models (the need for innovative lease contracts, availability of take-back systems, methods/channels to communicate with customers and end-users ...), significant for the effective recirculation of materials. Finally, the "Education, Acceptance and CE awareness" considers aspects neglected by current SBRS, even though Academia has identified the lack of CE skills, knowledge, and acceptance as one of the main barriers to circular approaches spread and implementation in the building sector [75, 76]. For example, DGNB addresses education and knowledge transfer concerns but is focused on users' relationship with nature and the ecological preservation of the building and its surroundings instead of spreading unpopular circularity concepts. SBTool mentions the control of "level of skills and knowledge of operating staff" among its criteria but, again, the metric does not refer to the variables involved in CE implementation in constructions.

The SBRS review also revealed which sustainability concerns are more relevant in terms of CE, and so, are more important when conceiving a CB. In this sense, Fig. 3 qualitatively shows the priority scale attributable to the identified criteria and the existing assessment methods involved in the proposed meta-protocol. That will be the base to subsequently define through an Analytic Hierarchy Process (AHP) the weighting system of the proposed assessment framework. Furthermore, BREEAM, LEED, DGNB, Green Globes, and Active House explicitly require a complete LCA, thus connecting circularity solutions to the quantification of environmental impact characterizing a building's lifecycle. Therefore, for quantitative evaluation of performances concerning "Life Cycle Environmental Impact Minimization", the meta-protocol suggests the calculation of environmental loads through a multi-LCA[10] [77, 78] and LCC for measuring the "Life Cycle Cost and Service Life Planning "criterion.

Micro circularity indicators demonstrated their usefulness in measuring and summarizing CE aspects strictly related to building material wasting. Therefore,

[10] Multi-LCA, approach to quantify the environmental impacts of products designed for multiple life cycles and to evaluate the flow of materials among the generated products.

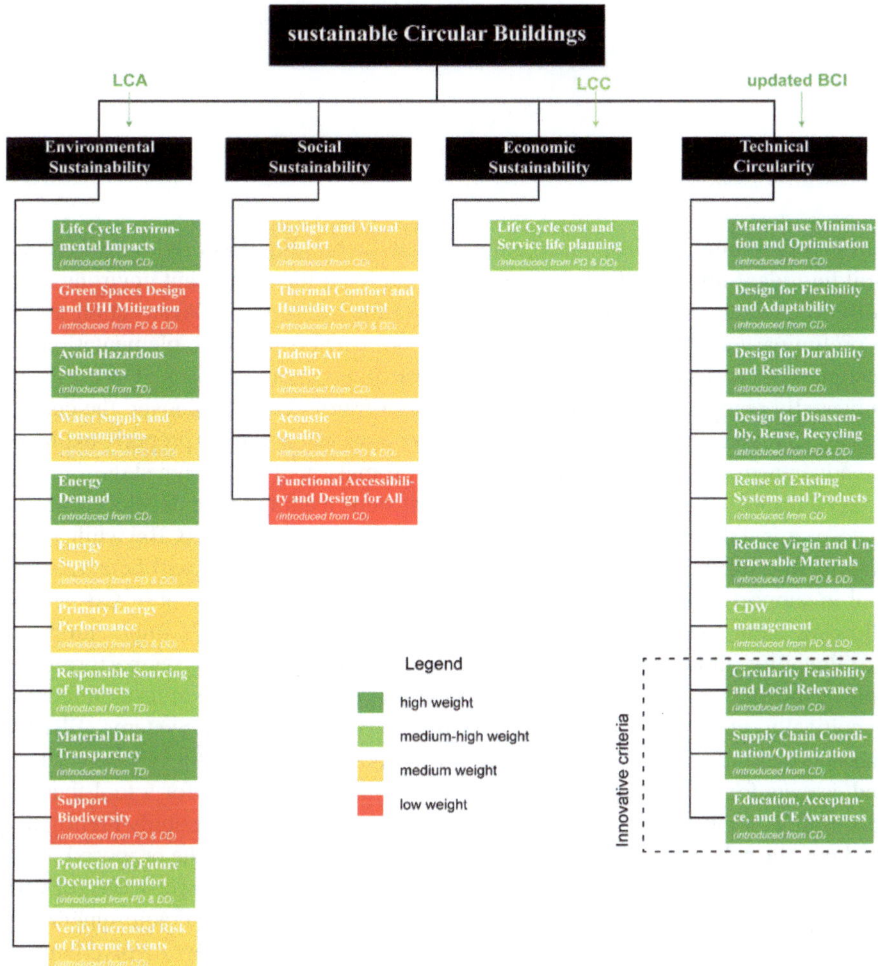

Fig. 3 Schematic representation of the comprehensive assessment framework at the base of the meta-protocol with the chromatic-priority scales association. CD refers to conceptual design, PD and DD refer to preliminary design and developed design, and TD means technical design (EN 16310:2013). Green criteria are more relevant ones (foreseen greater weight); red criteria are less relevant ones (foreseen less weight); yellow criteria are intermediated ones

the technical domain of CBs (the one concerning narrowing, slowing, and closing-loop strategies) will be quantitively assessed and resumed through an updated version of the BCI and adjusted circularity assessment frameworks. The other evaluations involved by the meta-protocol criteria will fill the gaps in these metrics thus providing a comprehensive assessment framework for buildings that considers circularity following sustainability concerns.

Finally, all relevant information coming from the involved evaluations will be then stored in a BIM environment (that will also be a relevant tool to streamline specific assessments) and eventually integrated into Materials/Products Passports, subsequently optimizing the previously mentioned "Organizational" domain.

3 Conclusions

Circular Economy and its conceptualization at the building level (with the Circular Building idea) are identified as possible solutions to the significant environmental pollution, waste generation, and resource depletion levels of the Construction sector. A preliminary analysis highlighted that relevant barriers to CB implementation are designers' and clients' distrust of "circular" constructions and the lack of uniform and universally recognized performance assessment frameworks. Moreover, the current state-of-the-art concerning building circularity assessment fails to properly explicate the advantages of CE strategies to key decision-makers of the design stage.

As a result, we presented fundamental first steps leading to the development of a meta-protocol that can encourage the design of Circular Buildings (CBs) while being compliant with sustainability key principles. The meta-protocol will be a valuable tool in supporting designers and clients during the decision-making of CB design, allowing them to understand and demonstrate the CE strategy's implications on sustainability dimensions. This paper illustrates the critical analyses performed to outline the meta-protocol's framework and report the first version of its architecture. Specifically, we reviewed existing Sustainable Building Rating Systems (SBRSs) and micro circular indicators for the Built Environment to detect the relevant criteria and aspects for CBs, thus setting up the theoretical background for a tool that can successfully support the planning of this new type of building. The study highlighted the lack of fundamental issues in SBRSs that, on the other hand, the proposed meta-protocol will better investigate: (i) context-specific logistic and infrastructural variables resulting from material-efficient use and re-circulation; (ii) requirements for effective data exchange and coordination between the stakeholders involved during the building lifecycle and beyond; (iii) remarks to improve CE knowledge and acceptance in the Built Environment. Regarding the micro circular indicators, the research underlines their importance in rationalizing and communicating in a user-friendly way the level of CE integration in buildings but also points out this metrics limitation, thus suggesting the coupling with broader evaluations. All these considerations led to the base structure of the proposed meta-protocol.

The following steps in the meta-protocol development are the elaboration of the defined criteria's indicators, benchmarks, and associated scoring system. The research project will then deepen the use of Information and Communication Technologies as BIM to optimize the required evaluations and for storing the main related results. Finally, the meta-protocol usability will be tested through the application to selected case studies. Further developments also include investigations about the required LCA completeness and most appropriate BIM-integrated LCA/LCC approaches at each design phase.

To conclude, we are working on a meta-protocol that will provide a comprehensive assessment framework for buildings with criteria to assess the three dimensions of sustainability and circularity. Most of the criteria will directly refer to or draw inspiration from ones in existing SBRSs and CE metrics for the built environment. Sustainability criteria will validate circularity features—which implementation level will be also measured by existing and more innovative metrics—during different stages of building design. A so-defined tool will support the creation of CBs while promoting CE patterns in the Construction sector.

Acknowledgements The authors acknowledge Minerva Stiftung (Fondazione Minerva) for supporting this research through a scholarship created as part of the 4R^{n}D circular construction hub project.

References

1. EU Commission, Energy performance of buildings directive. https://energy.ec.europa.eu/topics/energy-efficiency/energy-efficient-buildings/energy-performance-buildings-directive_en, last access 2024/01/08.
2. EU Commission, Buildings and construction. https://single-market-economy.ec.europa.eu/industry/sustainability/buildings-and-construction_en, last access 2024/01/08.
3. EU Commission: Roadmap to a Resource Efficient Europe (COM/2011/0571). Publications Office of the European Union, Luxembourg (2011).
4. EU Commission for Environment: A New Circular Economy Action Plan for a Cleaner and More Competitive Europe (COM/2020/98). Publications Office of the European Union, Luxembourg (2020).
5. Ellen MacArthur Foundation: Towards the Circular Economy. Economic and business rationale for an accelerated transition. Seacourt, Chicago (2013).
6. Circle Economy, DGBC, Metabolic, SGS Search: A Framework for Circular Buildings. Indicators for possible inclusion in BREEAM (2018).
7. Kanters, J.: Circular Building Design: An Analysis of Barriers and Drivers for a Circular Building Sector. Buildings 10 (4), 77 (2020).
8. Khadim, N., Agliata, R., Marino A., Thaheem, M.J.: Critical review of nano and micro-level building circularity indicators and frameworks. J. Clean. Prod. 357, 131859 (2022).
9. Pomponi, F., Moncaster, A.: Circular economy for the built environment: A research framework. J. Clean. Prod. 143, 710–718 (2017).
10. Antwi-Afari, P., Ng, S.T., Uzzal Hossain, Md.: A review of the circularity gap in the construction industry through scientometric analysis. J. Clean. Prod. 20, 126870 (2021).
11. MacLeamy, P.: Collaboration, integrated information and the project lifecycle in building design, construction and operation. WP-1202. The Construction Users Roundtable, (2004).
12. Reikea, D., Vermeulen, W.J., Witjes, S.: The circular economy: New or Refurbished as CE 3.0? Exploring Controversies in the Conceptualization of the Circular Economy through a Focus on History and Resource Value Retention Options. Resour Conserv Recycl 135, 246–264 (2018).
13. Korhonen, J., Honkasalo, A., Seppälä, J.: Circular Economy: The Concept and its Limitations. Ecol Econ 143, 37–46 (2018).
14. Sauvé, S., Bernard, S., Sloan, P.: Environmental sciences, sustainable development and circular economy: Alternative concepts for trans-disciplinary research. Environmental Development 17, 48–56 (2016).

15. Giorgi, S., Lavagna, M., Wang, K., Osmani, M., Liu, G.: Drivers and barriers towards circular economy in the building sector: Stakeholder interviews and analysis of five European countries policies and practices. J. Clean. Prod. 336, 130395 (2022).
16. Rigamonti L., Mancini, E.: Life cycle assessment and circularity indicators. Int J LCA 26, 1937–1942 (2021).
17. Norouzi, M., Chàfer, M., Cabeza, L.F., Jiménez, L., Boer, D.: Circular economy in the building and construction sector: A scientific evolution analysis. J. Build. Eng. 44, 102704 (2021).
18. Cagno, E., Negri, M., Neri, A., Giambone, M.: One Framework to Rule Them All: An Integrated, Multi-level and Scalable Performance Measurement Framework of Sustainability, Circular Economy and Industrial Symbiosis. Sustain. Prod. Consum. 35, 55–71 (2023).
19. Moreau, V., Sahakian, M., van Griethuysen, P., Vuille, F.: Coming Full Circle: Why Social and Institutional Dimensions Matter for the Circular Economy. J. Ind. Ecol. 21(3), 497–506 (2017).
20. Murray, A., Skene, K., Haynes, K.: The Circular Economy: An Interdisciplinary Exploration of the Concept and Application in a Global Context. J Bus Ethics 140, 369–380 (2017).
21. Dams, B., Maskell, D., Shea, A., Allen, S., Driesser, M., Kretschmann, T., Walker, P., Emmitt, S.: A circular construction evaluation framework to promote designing for disassembly and adaptability. J. Clean. Prod. 316(3), 128122 (2021).
22. Gonzàlez, A., Sendra, C., Herena, A., Rosquillas, M., Vaz, D.: Methodology to assess the circularity in building construction and refurbishment activities. RCR Advances 12, 200051 (2021).
23. Lei, H., Li, L., Yang, W., Bian, Y., Li, C.: An analytical review on application of life cycle assessment in circular economy for built environment. J. Build. Eng. 44, 103374 (2021).
24. Larsen, V.G., Tollin, N.P., Sattrup, A., Birkved, M.: What are the challenges in assessing circular economy for the built environment? A literature review on integrating LCA, LCC and S-LCA in life cycle sustainability assessment, LCSA. J. Build. Eng. 50, 104203 (2022).
25. Hasheminasab, H., Zolfani, S.H., Kharrazi, M., Streimikiene, D.: Combination of sustainability and circular economy to develop a cleaner building industry. Energy Build. 258, 111838 (2022).
26. Antonini, E., Boeri, A., Lauria, M., Giglio, F. Reversibility and Durability as Potential Indicators for Circular Building Technologies. Sustainability 12 (18), 7659 (2020).
27. Blessing, L.T. and Chakrabarti, A.: DRM, a Design Research Methodology. Springer, London (2009).
28. SBi, GXN: Guide to Sustainable Building Certifications. 1st ed. SBi and GXN (2018).
29. Vierra, S.: Green Building standards and certification systems (2016). https://www.wbdg.org/resources/green-building-standards-and-certification-systems, last access 2024/01/08.
30. Zimmermann, R.K., Skjelmose, O., Jensen, K.G., Jensen K.K., Birgisdottir, H.: Categorizing Building certification systems according to the definition of sustainable building. OP Conference Series: Materials Science and Engineering 471(8), 092060 (2019).
31. The international Active House Alliance Homepage. https://www.activehouse.info/, last access 2024/03/13.
32. The International Active House Alliance: The Active House Specification, 3rd ed. The International Active House Alliance (2020).
33. Di Santo, N., Guante Henriquez, L., Dotelli, G., Imperadori, M.: Holistic Approach for Assessing Buildings' Environmental Impact and User Comfort from Early Design: A Method Combining Life Cycle Assessment. BIM, and Active House Protocol Buildings 13(5), 1315 (2023).
34. BRE Group, What is BREEAM. https://bregroup.com/products/breeam/, last access 2024/03/13.
35. BRE Group Ltd: BREEAM International New Construction. Version 6.0. BRE (2021).
36. CASBEE Homepage. https://www.ibec.or.jp/CASBEE/english/, last access 2024/03/13.
37. Japan Sustainable Building Consortium (JSBC) and Institute for Building Environment and Energy Conservation (IBEC) CASBEE for Building (New Construction). Comprehensive Assessment System for Built Environment Efficiency. Technical Manual (2014 Edition). IBEC (2014).

38. DGNB Homepage. https://www.dgnb.de/en, last access 2024/03/13.
39. DGNB GmbH: DGNB System. New Construction, Buildings Criteria Set. Version 2023. Deutsche Gesellschaft für Nachhaltiges Bauen (German Sustainable Building Council) (2023).
40. DGNB GmbH: Circular Economy. Closing loops means being fit for the Future. Deutsche Gesellschaft für Nachhaltiges Bauen (German Sustainable Building Council) (2019).
41. Green Building Initiative, Green Globes Certification. https://thegbi.org/greenglobes/, last access 2024/03/13.
42. Green Building Initiative: New Construction 2021. Technical Reference Manual. Implementation of ANSI/GBI 01-2021: Green Globes Assessment. Version 1.0. Green Building Initiative (2022).
43. Green Rating for Integrated Habitat Assessment, GRIHA Rating. https://www.grihaindia.org/griha-rating, last access 2024/03/15.
44. GRIHA Council: GRIHA V.2019. Volume 1. Introduction to National Rating System—GRIHA. The Energy and Resources Institute (TERI) (2021).
45. U.S. Green Building Council, LEED rating system. https://www.usgbc.org/leed, last access 2024/03/15.
46. U.S. Green Building Council: LEED v4.1 Building Design and Construction. Getting starting guide for beta participants. U.S. Green Building Council (2023).
47. Minergie, Certificazione secondo il prodotto complementare ECO. https://www.minergie.ch/it/certificare/eco/, last access 2024/03/15.
48. Minergie, Ecobau: Catalogo dei requisiti per il complemento ECO (2023).
49. IISBE—International Initiative for a Sustainable Built Environment, CMStuning, SBTool and SNTool. https://www.iisbe.org/sbmethod, last access 2024/03/15.
50. SNBS, SNBS-Hochbau. https://www.snbs-hochbau.ch/standards/snbs-hochbau/, last access 2024/03/15.
51. Bundesamt für Energie (BFE), ICCCON AG, Netzwerk Nachhaltiges Bauen Schweiz (NNBS): Kriterienbeschrieb SNBS-HOCHBAU. Nutzungsarten Wohnen, Verwaltung, Bildungsbauten, Gewerbenutzung im Erdgeschoss. Version 2023.1. Netzwerk Nachhaltiges Bauen Schweiz (NNBS) (2023).
52. European Commission for Environment, Level(s), European framework for sustainable buildings. https://environment.ec.europa.eu/topics/circular-economy/levels_en, last access 2024/03/18.
53. Dodd N., Donatello S., Cordella M.: Level(s): A common EU framework of core sustainability indicators for office and residential buildings. User Manual 1: Introduction to the Level(s) common framework (Publication version 1.1). European Commission, Joint Research Centre (2021).
54. Kristensen, H.S., Mosgaard, M.A.: A review of micro level indicators for a circular economy e moving away from the three dimensions of sustainability. J. Clean. Prod. 243, 118531 (2020).
55. Verberne, J.J.H.: Building circularity indicators. Eindhoven University of Technology, Eindhoven (2016).
56. van Vliet, M.M.: Disassembling the steps towards building circularity redeveloping the building disassembly assessment method in the building circularity. Eindhoven University of Technology, Eindhoven (2018).
57. Geraedts, R.: FLEX 4.0, A Practical Instrument to Assess the Adaptive Capacity of Buildings. Energy Procedia 96, 568–579 (2016).
58. Fregonara, E., Giordano, R., Ferrando, D.G., Pattono, S.: Economic-Environmental Indicators to Support Investment Decisions: A Focus on the Buildings' End-of-Life Stage. Buildings 7(3), 65 (2017).
59. Sreekumar, N.: An integrated approach towards energy performance and circularity in buildings. Delft University of Technology, Delft (2019).
60. Foster, G., Kreinin, H., Stagl, S.: The future of circular environmental impact indicators for cultural heritage buildings in Europe. Environmental Sciences Europe 32, 141 (2020).
61. Valdebenito, G., Vásquez, V., Prieto, A.J., Alvial, J.: O Paradigma da Economia Circular na Conservação Patrimonial do sul do Chile. Arquitetura Revista 17, 73–89 (2021).

62. Finch, G., Marriage, G., Pelosi, A., Gjerde, M.: Building envelope systems for the circular economy; Evaluation parameters, current performance and key challenges. Sustain. Cities Soc. 64, 102561 (2021).
63. Abadi, M., Sammuneh, M.A.: Integrating Circular Economy and Constructability. In: Proceedings of the 36th Annual ARCOM Conference, pp. 516–525. Association of Researchers in Construction Management (ARCOM), Leeds (2020).
64. Navare, K., Muys, B., Vrancken, K.C., Van Acker, K.: Circular economy monitoring: How to make it apt for biological cycles? Resour Conserv Recycl 170, 105563 (2021).
65. Cottafava, D., Ritzen, M.: Circularity indicator for residential buildings: Addressing the gap between embodied impacts and design aspects. Resour Conserv Recycl 164, 105120 (2021).
66. Lei, H., Yang, W., Wang, W., Li, C.: A new method for probabilistic circular economy assessment of Buildings. J. Build. Eng. 57, 104875 (2022).
67. Heisel, F., Rau-Oberhuber, S.: Calculation and evaluation of circularity indicators for the built environment using the case studies of UMAR and Madaster. J. Clean. Prod. 243, 118482 (2019).
68. d.Hub Circular Building Toolkit Homepage. https://ce-toolkit.dhub.arup.com/framework 2023, last access 2024/03/22.
69. Santoro, A., van der Feen, L.: Measuring circular buildings—key considerations. World Business Council for Sustainable Development (WBCSD) (2022).
70. Fischer, A.: Building Value: A pathway to circular construction finance. Circle Economy, Sustainable Finance LAB (2019).
71. Geldermans, B., Jacobson, L.R.: Circular Material and Product Flows in Buildings. Delft University of Technology, Delft (2015).
72. Rahla, K.M., Bragança, L., Mateus, R.: Obstacles and barriers for measuring building's circularity. IOP Conference Series: Earth and Environmental Science 225, 012058 (2019).
73. Morel, J.C., Mesbah, A., Oggero, M., Walker, P.: Building houses with local materials: means to drastically reduce the environmental impact of construction. Building and Environment 36(10), 1119–1126 (2001).
74. Parece, S., Rato, V., Resende, R., Pinto, P., Stellacci, S.: A Methodology to Qualitatively Select Upcycled Building Materials from Urban and Industrial Waste. Sustainability 14(6), 3430 (2022).
75. Tokazhanov, G., Galiyev, O., Lukyanenko, A., Nauyryzbay, A., Ismagulov, R., Durdyev, S., Turkyilmaz, A., Karaca, F. Circularity assessment tool development for construction projects in emerging economies. J. Clean. Prod. 362, 132293 (2022).
76. Bilal, M., Ahmad Khan, K.I., Thaheem, M.J., Nasir, A.R. Current state and barriers to the circular economy in the building sector: Towards a mitigation framework. J. Clean. Prod. 276, 123250 (2020).
77. Guante Henriquez, L., Lavagna, M., Frontini, F. and Imperadori, M.: LCA to support the integration of Reuse strategies during building design: advancements and gaps. In: XVII Convegno della Associazione Rete Italiana LCA. 30 anni di Life Cycle Assessment, sviluppi metodologici e applicativi. pp. 686–695. Associazione Rete Italiana LCA, Milano (2023).
78. Suhariyanto, T.T., Wahab, D.A., Rahman, M.N.A.: Multi-Life Cycle Assessment for sustainable products: A systematic review. J. Clean. Prod. 165, 677–696 (2017).

Taking Off Towards Circularity: Measuring Circular Economy Practices in Aviation and Beyond

Christine Grimm

Abstract This study addresses the challenge of measuring circularity within procurement-to-waste system boundaries, where existing methodologies were deemed inadequate due to their complexity or limited scope. To fill this gap, we propose an extension of the Circularity Mass Utilisation (CMU) Index, currently employed at the EU level, by incorporating circular inflow to better account for procurement decisions. The research culminates in the following formula: CMU Extended = (Circular Inflow + Circular Outflow)/(2 * Total Material Use). Successful testing with SWISS International Air Lines Ltd., using single-use in-flight drinking cups as a case study, demonstrates its practical relevance in the industry and suggests that the approach is likely to be applicable across various sectors beyond the airline industry.

Keywords Circular economy · Measuring circular economy · Indicator circular economy · Sustainability · Procurement · Waste · Aviation industry

1 Introduction

As the circular economy gains traction, businesses and organizations play a crucial role in advancing sustainable and resource-efficient practices. Measuring circularity is vital for companies to evaluate and track their progress in adopting circular economy principles (e.g., [1–3]). Metrics at the company level can highlight areas for integrating circular practices, such as enhancing product durability and establishing recycling and reuse chains. Understanding circularity performance helps companies

C. Grimm (✉)
Institute for Innovation and Technology Management, Lucerne University of Applied Sciences and Arts, Horw, Switzerland
e-mail: Christine.Grimm@hslu.ch

A. Podleisek et al. (eds.), *Advances in Circular Economy: Bridging Research and Practice*, Springer Proceedings in Earth and Environmental Sciences,
https://doi.org/10.1007/978-3-032-21547-5_8

align with broader sustainability goals, improve competitiveness, reduce environmental impacts, and meet regulatory requirements [4]. However, measuring circularity remains complex, requiring standardized methods and comprehensive data collection.

This study addresses the challenge of measuring circularity within the procurement-to-waste system boundaries, using the example of single-use in-flight drinking cups at SWISS International Air Lines Ltd. (SWISS). In 2019, SWISS served 22.4 million plastic cups for cold beverages, all of which went to waste (SWISS internal documentation). Since the company does not manufacture these cups or manage their waste, its influence on the value chain is limited, a common scenario for many inflight products.

The goal is to develop an index suitable for scenarios where the company acts as a B2B buyer rather than a producer. This index should be applicable to various products, using the company's proprietary economic data to ensure accuracy, validity, and security. It should also adapt to changing conditions, including shifts in political regulations.

Following this introduction, section two outlines the research design, while sections three and four present a literature review and evaluate existing market tools. Section five introduces and discusses our newly developed circularity metric, addressing the gaps identified in previous sections and forming the basis for the study's conclusions.

2 Research Method

The study used an exploratory research approach to address the challenge of measuring circular economy practices. The methodology included a comprehensive literature review, market analysis of existing tools, stakeholder interviews, and the development of a new metric.

A systematic literature review was conducted using Google Scholar and the research team's library to identify and analyze key papers on circularity measurement, establishing a solid theoretical foundation. The review focused on keywords like "measuring circular economy" and "measuring recycling," and screened for papers with mathematical and systematic approaches, particularly those relevant to the case study's procurement-to-waste system boundaries.

Following this, a market analysis evaluated current tools for measuring circular economy practices. This involved listing tools available in the industry and assessing their effectiveness, applicability, and limitations. Tools were analyzed to determine their suitability for the specific procurement-to-waste system boundaries of the case study. Additionally, exploratory semi-structured online interviews with selected companies and email exchanges provided further insights. Interviews with Swiss government and Kanton Zürich representatives offered additional perspectives on measuring circularity.

Based on the gaps identified, the research team developed a new metric, CMU Extended, by refining and validating an existing circularity metric to address these gaps effectively.

3 State of the Art: Literature

The literature review focused on methodologies within the procurement-to-waste system, emphasizing buyer perspective and prioritizing directly applicable, quantitative solutions over conceptual discourse. The study examined the work of Parchomenko et al. [5], who clustered 63 circular economy metrics into three categories: resource-efficiency, materials stocks and flows, and product-centric. Following the question of relevance to context, we focused on metrics related to materials stocks and flows and product-centric metrics. Within the materials stocks and flows cluster, nine metrics were identified, primarily addressing post-procurement and waste phases, with some covering production: Haupt and Zschokke [6], Jones et al. [7], Park and Chertow [8], Ma et al. [9], Zhang et al. [10], Krausmann et al. [11], Laner et al. [12], Schiller et al. [13], Kasulaitis et al. [14]. Whilst nine out of nine addressed the waste perspective, notably, none addressed the procurement phase. In the product centric cluster, we find a similar picture, investigation the work of Dahmus and Gutowski [15], Wurster and Ladu [16], Evans and Bocken [17], Nelen et al. [18], Franklin-Johnson et al. [19], Singh and Ordoñez [20], Akanbi et al. [21], Schipper et al. [22], Figge et al. [1]. Eight out of nine papers address the waste perspective, none considered the procurement phase.

In the quest for procurement insights, over 40 additional papers were screened, but only three were found to look at the procurement phase. Bai et al. [23] proposed a complex decision-making method for circularity supplier evaluation, which, despite its intriguing premise, is impractical for the context given, due to its complexity. Aktin and Gergin [24] emphasized the need to consider environmental, economic, and social dimensions in supplier evaluation, however, relied on questionnaire responses, raising validity concerns. Sheu et al. [25] introduced a green-supply chain model more suited to manufacturing than service providers like SWISS Airlines and is thus also not suitable for the case.

Regarding waste management, discussions centered on the balance between methodological detail and data feasibility. Park and Chertow [8] introduced the Reuse Potential Indicator (RPI), Anastasiades et al. [26] developed a Circular Construction Indicator framework incorporating a reuse potential indicator. On EU level, we find the Circular Material Use Indicator (CMU) published by Eurostat [27] wherein two variables—recycled material and total material used—are juxtaposed in a ratio. The approach quantifies the proportion of material recycled and reintegrated into the economy relative to the total material utilized. The indicator relies on mandatory reporting from all EU Member States, alongside aggregated member state data [27]. At the national level, the CMU rate is further subdivided by material categories [28].

Pires and Martinho [29] introduced a waste hierarchy index (WHI) based on reuse and upcycling, excluding incineration as a circular operation. Other discussions reflected on more detailed waste classifications and proposed indicators like the Recyclability Benefit Rate and the Recycled Content Benefit Rate. Nelen et al. [18] aimed to develop multidimensional indicators for recycling of waste of electrical and electronic equipment, introducing innovative environmental impact avoidance factors.

Interesting and standing out, Figge et al. [30] proposed a portfolio theory of the circular economy, viewing circularity as a spectrum of configurations with varying environmental and economic implications. The authors suggested that individual company actions can influence group behavior, potentially shaping overall circularity. This group-level perspective offers valuable insights for future research.

4 State of the Art: Tool Market

Initially, market research was conducted to identify and assess the current tools available. This involved an analysis of the market, supplemented by reflections found in the literature by Sacco et al. [31], Vinante et al. [3] and Valls-Val et al. [2]. Based on these sources, the following tools were selected for further evaluation:

1. Circulytics
2. MCI/material circularity indicator
3. CTI
4. Acodea
5. CEEI
6. CircularTrans
7. Inedit
8. TECNUN
9. CircularityCalculator
10. KMPG

Following the identification of current tools, an assessment was conducted based on the succeeding criteria: active users, past users, Europe/global, costs, language, quantitative, qualitative, materials, water, waste, energy, finance, on the market since, likelihood to stay on the market, potential policy impact, product level assessment, company level assessment, education of user. In search for a tool which would allow an industry wide comparison on product level now and in the future, priority has been given to the categories:

- Active/Passive Users: Indicator for market acceptance
- Costs: Value for money
- On the market since: Indicator for market acceptance
- Likelihood to stay on the market: Indicator for market acceptance
- Policy impact: Indicator to evaluate if the tool reflects current EU/Swiss regulations and will do so in the future

- Product level assessment: Assessment on product level

Initially, ten tools were identified for evaluation, but five were excluded due to access restrictions, unavailability, or language barriers. KPMG's tool, which is based on the CTI tool, was also eliminated, leaving four tools for investigation. Analysing the different tools with the categories described above, we came to the—for us surprising—conclusion, that actually none of these tools is well-suited to our case study. While Circulytics is robust and has potential market impact, its scope goes beyond our project's objectives by assessing the entire company. The MCI tool, while simple, lacks applicability for products without longevity data and is difficult to modify. The CTI tool faces challenges due to SWISS's lack of ownership over necessary data. Although the Circularity Calculator has a well-designed interface and comprehensive documentation, its focus on early-stage product design guidance and differentiating on various loop types hinder its usability for our project.

Overall, the assessment reveals a lack of solutions tailored to companies in a procurement position without ownership of the product's value chain. Additional common challenges which we found across all tools include: a lack of cost-revenue structures, data governance issues, reliance on product suppliers, and standalone functionality without integration with existing tools. These findings align with existing literature, which highlights the absence of standardized approaches and the challenge of balancing analysis depth with data collection requirements.

In assessing the tool's applicability, we also considered the relationship between the tool provider and political institutions, acknowledging that political regulations significantly influence corporate actions. Our investigation, consisting of studying legislative publications as well as interviews with the county Zürich and the Swiss government reveals a lack of standardized metrics or mandatory frameworks at the Swiss national level, its counties, or the EU. While existing EU measures address various aspects of circularity, they primarily operate at the country and EU levels, focusing on material flows overall, not on company level. Discussions with the Swiss government suggest that mandating circularity metrics at the company level within the EU and Switzerland is considered to as unlikely due to enforcement challenges.

5 Development of a Circularity Factor

Due to the lack of suitable approaches in existing literature, tools, or regulations to address procurement-to-waste decisions in a business-to-business (B2B) context, in an iterative process we re-evaluate all available information. Through these iterations, including an analysis of measures by the Swiss government and the European Union (EU), we re-visited a familiar metric used at the country and EU level: the Circular Material Use Indicator (CMU) [27]. As described in the literature section, this approach bears resemblance to the basic metric described in waste management literature, wherein two variables—recycled material and total material used—are

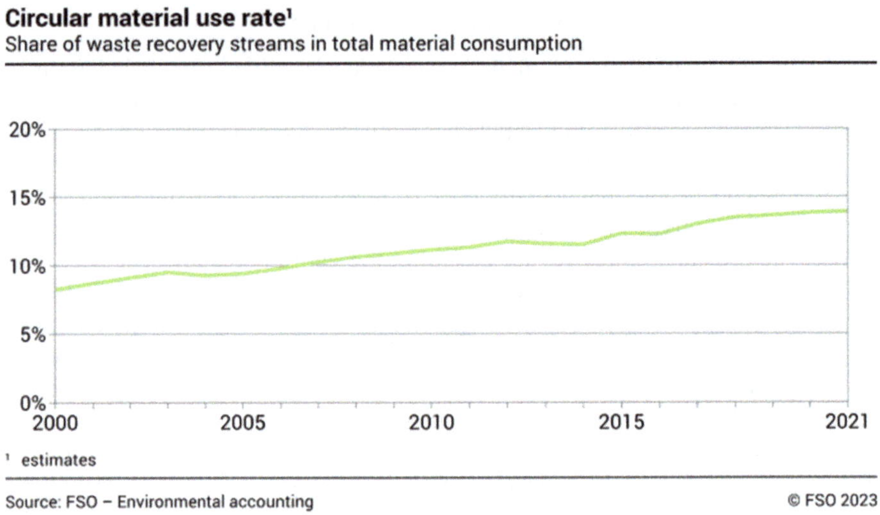

Fig. 1 CMU displayed and published by the Swiss government [28]

juxtaposed in a ratio leading to the following equation: Circular Material Use Indicator (CMU) = Circular Use of Material (U)/Overall Material Use (M). An example published by the Swiss government on country level is shown in the Fig. 1.

The CMU holds significance on various fronts. It is established both within SWISS's home country and in target markets like the European Union (EU) and associated countries, making it familiar to political entities and the informed public. Its communication simplicity as a percentage enhances accessibility. Furthermore, data availability is ensured as values for the necessary variables are compulsorily reported by each EU and associated government, including Switzerland.

However, applying this metric within a country context faces naturally also limitations. Firstly, the reliability of the indicator depends on the quality of data provided by member states, which may vary, compromising accuracy. Secondly, the indicator only includes materials officially designated as waste, excluding on-site loops like internal recycling within companies. Thirdly, focusing solely on the disposal phase introduces challenges such as complexity versus granularity in analysis, the absence of reflection on the procurement phase and the ignorance of any other type of looping than recycling. Will these limitations persist if applied at the company level? Could including the procurement phase by introducing an additional variable enhance the comprehensiveness of this otherwise straightforward yet potent metric? These questions require further exploration and will be addressed in subsequent discussions within this paper.

5.1 *Transfer of an Existing Approach*

We embarked on a quest to address two fundamental questions:

1. Is it feasible to tailor the existing CMU to assess circularity at the company or product level, with a specific emphasis on material outflow?
2. Is it feasible to modify the existing CMU to incorporate both the procurement and waste decision phases?

In pursuit of the first question, we conducted two tests, using the case of SWISS and the challenge to measure circularity in the context of the plastic cups for cold beverages which are used in flight, as described in the introduction to this paper. The details of this test are delineated below.

5.2 *Testing the CMU on Company Level*

Example 1: We have two products: the plastic glass large and the r-PET Cup. We assume that the plastic glass large as well as the r-PET Cup are not going into any circular loop and that we have 100 units in total:

$$\begin{aligned}
&\text{U(circular use of material)} = 0\text{(nothing is looped)}\\
&\text{M(total material)} = 100\\
&\text{CMU} = 0/100\\
&\text{CMU} = 0\%
\end{aligned}$$

Example 2: We assume that 80 units of the plastic glass large as well as 80 units of the r-PET Cup are going into recycling and that we have 100 units in total:

$$\begin{aligned}
&\text{U(circular use of material)} = 80\text{(80 units are looped)}\\
&\text{M(total material)} = 100\\
&\text{CMU} = 80/100\\
&\text{CMU} = 80\%
\end{aligned}$$

What we can see is that in its simplicity to calculate, we can translate the approach to the current case and with this apply it on company and even product level. The figure illustrates the advantages and disadvantages if we were to follow this approach and use this equation for the case of SWISS (Figs. 2 and 3).

Confirming the applicability of the CMU on company and product level, we turned our attention to addressing the second question:

1. Is it feasible to modify the existing CMU to incorporate both the procurement and waste decision phases?

Cup Type	Circular Use of Material	Overall Material Use	CMU
Plastic Glass Large (LX31.020)	0	100	0%
r-PET Cup	0	100	0%
	Nothing recycled		

Cup Type	Circular Use of Material	Overall Material Use	CMU
Plastic Glass Large (LX31.020)	80	100	80%
r-PET Cup	80	100	80%
	80 untis will be recycled		

Fig. 2 Testing of CMU on company level

Pro	Contra
Allows quick assessment of circularity within in-house (available) data	Applied only on country level until now
Easy to communicate	Does not explicitly reflect loops
Easy to measure	Does not reflect cost/revenue of action
Data ownership	Not industry standard (there is none)
	Does not reflect procurement decisions

Fig. 3 Summary of advantages and disadvantages of CMU application on company level

5.3 Further Development of the CMU

To reflect the procurement decision, we modified the model such that it shows in- and outflows of material as displayed in the equation below:

$$CMU\ Extended = \frac{Circular\ Inflow + Circular\ Outflow}{2 * Total\ Material\ Flow}$$

Definition Outflow.

The outflow aligns with the original definition of Circular Use of Material (U). In the existing CMU equation at the national level, outflow encompasses all waste declared as such and subsequently recycled - on site looping as well as energy recovery options are not included [32].

Definition Inflow.

We propose enhancing the existing CMU by incorporating this variable to account for circular procurement decisions. Circular inflow is represented by the proportion of recycled material in a purchased product (as illustrated in the example below).

Total Material Flow.

Cup Type	Circular Use of Material (inflow)	Cirucular Use of Material (outflow)	Total Material	CMU
Large Plastic Glass (LX31.020)	0	0	100	0%
r-PET Cup	50	0	100	25%
	Contains 50% recycled material	Nothing recycled		
Example 2				
Large Plastic Glass (LX31.020)	0	80	100	40%
r-PET Cup	50	80	100	65%
	Contains 50% recycled material	80% will be recycled		

Fig. 4 Application of CMU extended to the cup case

The use of total material as the denominator aligns with the original definition. However, with the addition of the inflow variable, it is necessary to multiply the denominator by two to maintain equal weighting of both variables (normalization).

Testing.

Having extended the equation to include the procurement decision, we tested it on the example of cups as displayed below (Fig. 4).

In the example two, re-PET cup, we can see:

Circular Inflow (procurement): 100 kg cups contain 50 kg recycling material.

Circular Outflow: 80 kg of the 100 kg are recycled after being declared as waste.

Total Material: 2*100 kg.

CMU Extended: (50 kg + 80 kg)/2*100 kg = 0.65 = 65%

The example demonstrates that the CMU Extended formula successfully integrates procurement decisions while preserving the advantages of the original CMU method. Below, we offer a table summarizing the pros and cons, now enriched with insights from the modified equation (Fig. 5).

In italics, we emphasized the additional benefits of extending the model while eliminating a key disadvantage: the approach now effectively incorporates procurement decisions. This addresses a critical challenge in both academic literature and industry practice by identifying an indicator that delineates the procurement-to-waste system boundaries with a reasonable degree of effort.

Pro	Contra
Allows quick assessment of circularity within in-house (available) data	Applied only on country level until now
Easy to communicate	Does not explicitly reflect loops
Easy to measure	Does not reflect cost/revenue of action
Data ownership	Not industry standard (there is none)
Connects inflow with outflow -> reflects procurement decision	~~Does not reflect procurement decisions~~
Can easily be adapted to more detailed analysis	

Fig. 5 CMU-extended pro/contra

6 Conclusion

The analysis highlights a substantial gap in the existing literature, tools, and policies concerning circularity metrics within the procurement-to-waste system boundaries. While Parchomenko et al. [5] conducted a comprehensive literature review, none of the studies examined adequately addressed the procurement phase. A subsequent search identified three relevant papers: Bai et al. [23], Aktin and Gergin [24], and Sheu et al. [25]. Although Bai et al. [23] provided important insights, its complexity rendered it impractical for broad application. Aktin and Gergin [24] offered a questionnaire-based approach, yet concerns were raised regarding its credibility. Sheu et al. [25] focused on a different sector, limiting the applicability of their findings. Existing methodologies, such as the Circular Material Use (CMU) approach [27], are primarily concerned with waste management, marking the end of our system boundaries. However, comprehensive solutions encompassing the entire procurement-to-waste phase remain absent.

An examination of the market reveals that few tools surveyed focus specifically on circularity at the product level, and those that do encounter significant challenges related to data availability, system boundaries, and complexity. From a regulatory perspective, discussions with representatives from the Canton of Zürich and the Swiss government indicated no immediate plans to introduce specific regulations addressing circularity metrics in the airline industry or other sectors. Similarly, an analysis of EU legislation found no evidence that the issue of circularity metrics at the company level will be addressed in the near future.

In light of the inadequacy of existing methods suggested in the literature, the absence of suitable tools, and the lack of guidance from regulatory bodies, our iterative analysis of the data led to the adaptation of the Circularity Mass Utilization (CMU) Index, which is currently applied at the national level in Switzerland and the EU. Originally designed for macro-level applications, the CMU was hypothesized to be adaptable to both the company and product levels. Initial testing validated this hypothesis, prompting further refinement of the CMU to include the procurement variable, resulting in the CMU Extended formula: CMU Extended = (Circular Inflow + Circular Outflow)/(2 * Total Material Use). This revised formula incorporates both circular inflow (procurement) and circular outflow (waste management) in relation to total material use.

The successful application of this extended equation, using data from SWISS International Air Lines Ltd., demonstrated its efficacy in assessing circularity within the system boundaries from procurement to waste at the product level. Based on the results of this case study, it appears plausible that the indicator can be applied to products beyond the airline industry, within the specified system boundaries. Therefore, the CMU Extended may offer a valuable solution for both industry stakeholders and policymakers, providing a viable means of measuring circular economy efforts with reasonable economic effort.

For advancing company-level progress regarding sustainability, harnessing the power of CMU Extended in conjunction with factors such as life cycle assessment

(LCA) outcomes, financial ramifications, and CO_2 emissions data has the potential to offer a holistic perspective, equipping decision-makers with valuable insights for well-informed decision-making.

Acknowledgements This work would not have been possible without the invaluable contributions of Matthias Benz, a visionary mathematician. Special gratitude is extended to Dr. Laura Braun, Trisha Baumeler, and Louis Schober from SWISS Airlines Ltd. for presenting us with an exceptional research opportunity.

References

1. Figge, F. et al. (2018) 'Longevity and Circularity as Indicators of Eco-Efficient Resource Use in the Circular Economy', Ecological Economics, 150, pp. 297–306. https://doi.org/10.1016/j.ecolecon.2018.04.030.
2. Valls-Val, K., Ibáñez-Forés, V. and Bovea, M. D. (2022) 'How can organisations measure their level of circularity? A review of available tools', Journal of Cleaner Production, 354. https://doi.org/10.1016/j.jclepro.2022.131679.
3. Vinante, C. et al. (2021) 'Circular economy metrics: Literature review and company-level classification framework', Journal of Cleaner Production. Elsevier Ltd. https://doi.org/10.1016/j.jclepro.2020.125090.
4. Kulakovskaya, A., Knoeri, C., Radke, F., and Blum, N. U. (2023) Measuring the economic impacts of a circular economy: an evaluation of indicators abstract Circular Economy and Sustainability, 3(2), pp. 657–692. https://doi.org/10.1007/s43615-022-00190-w. https://link.springer.com/article/10.1007/s43615-022-00190-w
5. Parchomenko, A. et al. (2019) 'Measuring the circular economy—A Multiple Correspondence Analysis of 63 metrics', Journal of Cleaner Production, 210, pp. 200–216. https://doi.org/10.1016/j.jclepro.2018.10.357.
6. Haupt, M. and Zschokke, M. (2017) 'How can LCA support the circular economy?—63rd discussion forum on life cycle assessment, Zurich, Switzerland, November 30, 2016', International Journal of Life Cycle Assessment, 22(5), pp. 832–837. https://doi.org/10.1007/s11367-017-1267-1.
7. Jones, P. T. et al. (2013) 'Enhanced Landfill Mining in view of multiple resource recovery: a critical review', Journal of Cleaner Production, 55, pp. 45–55. https://doi.org/10.1016/J.JCLEPRO.2012.05.021.
8. Park, J. Y. and Chertow, M. R. (2014) 'Establishing and testing the "reuse potential" indicator for managing wastes as resources', Journal of Environmental Management, 137, pp. 45–53. https://doi.org/10.1016/J.JENVMAN.2013.11.053.
9. Ma, S. et al. (2015) 'A case study of a phosphorus chemical firm's application of resource efficiency and eco-efficiency in industrial metabolism under circular economy', Journal of Cleaner Production, 87(1), pp. 839–849. https://doi.org/10.1016/J.JCLEPRO.2014.10.059.
10. Zhang, L., Cai, Z., Yang J., Yuan, Z., Chen, Y. 'The future of copper in China—A perspective based on analysis of copper flows and stocks', Science of the Total Environment, 536. https://doi.org/10.1016/j.scitotenv.2015.07.021.
11. Krausmann, F. et al. (2017) 'Global socioeconomic material stocks rise 23-fold over the 20th century and require half of annual resource use', Proceedings of the National Academy of Sciences of the United States of America, 114(8), pp. 1880–1885. https://doi.org/10.1073/pnas.1613773114.
12. Laner, D., Zoboli, O. and Rechberger, H. (2017) 'Statistical entropy analysis to evaluate resource efficiency: Phosphorus use in Austria', Ecological Indicators, 83, pp. 232–242. https://doi.org/10.1016/J.ECOLIND.2017.07.060.

13. Schiller, G., Gruhler, K. and Ortlepp, R. (2017) 'Continuous Material Flow Analysis Approach for Bulk Nonmetallic Mineral Building Materials Applied to the German Building Sector', Journal of Industrial Ecology, 21(3), pp. 673–688. https://doi.org/10.1111/jiec.12595.
14. Kasulaitis, B. V, Babbitt, C. W. and Krock, A. K. (2019) 'Dematerialization and the Circular Economy: Comparing Strategies to Reduce Material Impacts of the Consumer Electronic Product Ecosystem', Journal of Industrial Ecology, 23(1), pp. 119–132. https://doi.org/10.1111/jiec.12756.
15. Dahmus, J. B. and Gutowski, T. G. (2007) 'What gets recycled: An information theory based model for product recycling', Environmental Science and Technology, 41(21), pp. 7543–7550. https://doi.org/10.1021/es062254b.
16. Wurster, S. and Ladu, L. (2022) 'Triple-C: A Tridimensional Sustainability-Oriented Indicator for Assessing Product Circularity in Public Procurement', Sustainability (Switzerland), 14(21). https://doi.org/10.3390/su142113936.
17. Evans, J. L. and Bocken, N. (2014) 'A tool for manufacturers to find opportunity in the circular economy'. https://www.researchgate.net/publication/278032838.
18. Nelen, D. et al. (2014) 'A multidimensional indicator set to assess the benefits of WEEE material recycling', Journal of Cleaner Production, 83(2014), pp. 305–316. https://doi.org/10.1016/j.jclepro.2014.06.094.
19. Franklin-Johnson, E., Figge, F. and Canning, L. (2016) 'Resource duration as a managerial indicator for Circular Economy performance', Journal of Cleaner Production, 133, pp. 589–598. https://doi.org/10.1016/J.JCLEPRO.2016.05.023.
20. Singh, J. and Ordoñez, I. (2016) 'Resource recovery from post-consumer waste: important lessons for the upcoming circular economy', Journal of Cleaner Production, 134, pp. 342–353. https://doi.org/10.1016/J.JCLEPRO.2015.12.020.
21. Akanbi, L. A. et al. (2018) 'Salvaging building materials in a circular economy: A BIM-based whole-life performance estimator', Resources, Conservation and Recycling, 129, pp. 175–186. https://doi.org/10.1016/J.RESCONREC.2017.10.026.
22. Schipper, B. W. et al. (2018) 'Estimating global copper demand until 2100 with regression and stock dynamics', Resources, Conservation and Recycling, 132, pp. 28–36. https://doi.org/10.1016/J.RESCONREC.2018.01.004.
23. Chunguang, Bai Qingyun, Zhu Joseph, Sarkis (2024) Circular economy and circularity supplier selection: a fuzzy group decision approach International Journal of Production Research, 62(7), pp. 2307–2330. https://doi.org/10.1080/00207543.2022.2037779
24. Tülin, Aktin Zeynep, Gergin (2016) Mathematical modelling of sustainable procurement strategies: three case studies Journal of Cleaner Production, pp. 113767–113780. https://doi.org/10.1016/j.jclepro.2015.11.057
25. Jiuh-Biing, Sheu Yi-Hwa, Chou Chun-Chia, Hu (2005) An integrated logistics operational model for green-supply chain management Transportation Research Part E: Logistics and Transportation Review, 41(4), pp. 287–313. https://doi.org/10.1016/j.tre.2004.07.001
26. Kostas, Anastasiades Johan, Blom Amaryllis, Audenaert (2023) Circular Construction Indicator: Assessing Circularity in the Design Construction and End-of-Life Phase Recycling, 8(2), pp. 29. https://doi.org/10.3390/recycling8020029
27. Eurostat (2018) Circular material use rate, Statistical Office of the European Communities. http://ec.europa.eu/eurostat/tgm/table.do?tab=table&init=1&plugin=1&language=en&pcode=cei_srm030.
28. Federal Statistik Office Switzerland (2024) Circular material use rate—Share of waste recovery streams in total material consumption. https://www.bfs.admin.ch/bfs/en/home/statistics/catalogues-databases.assetdetail.25866877.html.
29. Ana, Pires Graça, Martinho (2019) Waste hierarchy index for circular economy in waste management Waste Management, pp. 95298–95305. https://doi.org/10.1016/j.wasman.2019.06.014
30. Figge, F., Thorpe, A. S. and Manzhynski, S. (2021) 'Between you and I: A portfolio theory of the circular economy', Ecological Economics, 190, p. 107190. https://doi.org/10.1016/j.ecolecon.2021.107190.

31. Sacco, P. et al. (2021) 'Circular economy at the firm level: A new tool for assessing maturity and circularity', Sustainability (Switzerland), 13(9). https://doi.org/10.3390/su13095288.
32. Official Journal of the European Union (2008) DIRECTIVE 2008/98/EC OF THE EUROPEAN PARLIAMENT AND OF THE COUNCIL. https://eur-lex.europa.eu/legal-content/EN/TXT/HTML/?uri=CELEX:32008L0098.

Folded Steel and Cement-Free Concrete Composite Systems: Joint Design Towards Design for Disassembly

Jay Renée Thalmann, Yasaman Yavaribajestani, Christian Stocker, Peter Kobel, Samuel R. Garcia, Valentino Vigneri, Andreas Taras, and Patric Fischli-Boson

Abstract Despite the widespread use of recycled steel in construction, the industry still has difficulty meeting the net zero CO_2 emission goals due to the significant gray energy losses during building demolition and recycling. Thus more emphasis has been put on re-using construction material with the help of design for disassembly. The application of visible steel in residential construction (especially in Switzerland) has been limited due to fire protection regulations and the challenges associated with practices to address this matter. Concrete-steel hybrid structures, commonly used for fireproofing steel structures, come with high CO_2 emissions and lack easy dismantling, rendering them not compatible with sustainability goals. Furthermore, with hybrid steel-wood systems, the composite action is often not well developed, leading to inefficient use of material. The authors have introduced a demountable folded steel and cement-free concrete composite system that meets the fireproofing requirements and reduces the gray energy and raw material. The additively joined construction facilitates the later reuse of the elements. The structural system features small-scale steel beams and columns designed for residential buildings, using standardized components. The system's modularity supports architectural flexibility and adaptability for new constructions and extensions. This paper focuses on the details and development of joints and connections within this system that allow easy and efficient assembly and disassembly process, promoting circular economy in the construction and therefore playing a crucial role in improving the sustainability aspects and structural integrity of the hybrid steel structures.

Keywords Design for disassembly · Circular construction · Composite structure · Joint design · Sustainable construction · Reuse · Steel construction · Hybrid cross-section · Cement-free concrete · Partially-encased composite beams · Additive construction

---

J. R. Thalmann (✉) · Y. Yavaribajestani · C. Stocker · P. Kobel · S. R. Garcia · P. Fischli-Boson
Zürich University of Applied Science (ZHAW), Winterthur, Switzerland
e-mail: jt@ressegattithalmann.ch

V. Vigneri · A. Taras
Swiss Federal Institute of Technology in Zürich (ETH), Zürich, Switzerland

A. Podleisek et al. (eds.), *Advances in Circular Economy: Bridging Research and Practice*, Springer Proceedings in Earth and Environmental Sciences,
https://doi.org/10.1007/978-3-032-21547-5_9

1 Introduction

The steel construction industry, despite its widespread use of 99% recycled steel, faces challenges to meet the demand for sustainable building solutions. Given the 2050 world's climate-neutral goals [1], there is a pressing need for more sustainable measures within the construction methods. In this context, a significant challenge lies in the management of gray energy that is often lost during the demolition and recycling process of buildings. Although a substantial portion of the building's demolition waste will be recycled, much of the gray energy that was originally required to produce the building materials is lost in the recycling process. Therefore, the Swiss Federal Office for the Environment (FOEN) promotes reuse in the construction process to reduce the loss of gray energy and raw materials [2], requiring solutions that take into account the entire life cycle of the structure. This includes topics such as the possibility of preservation through conversion and densification, as well as the reuse of building materials associated with demolition. As shown in Fig. 1, adopting circular reuse practices, presents the most efficient and immediate approach to mitigating greenhouse gas emission production, when considering all stages of the linear construction process (A1-C4). [3]

Furthermore, in Switzerland, steel is not commonly used as a (visible) structural element in residential building construction, mainly due to fire protection regulations. Common practices, used in office buildings, like fire protection coatings and sprinkler systems, are maintenance-intensive and thus, not practical in residential settings.

There have been a lot of different efforts to improve the fire performance of steel structures. The combined use of structural steel and concrete is already an established construction method for fulfilling fire protection requirements and improving load-bearing capacity through composite action, outlined in standards such as the Swiss Society of Architects and Engineers SIA 264 [4] and SIA 264/1 [5]. However, from a circularity and sustainability aspects, this method presents some drawbacks. Partially encased composite steel-concrete beams [6, 7] present challenges in sustainability due to the difficulty of separating the bond at the end of their life cycle and the high CO_2 emissions from concrete production, which have an unfavorable impact on sustainability, disassembly, recycling, and circularity. Whilst steel beams could also be encased in timber elements for fire protection purposes, this solution does not rely on the composite action, and the load-bearing resistance of each of the materials is considered separately. For example, a primary support structure is designed as pure steel construction, on which the components of the secondary support structure (made of timber) are placed. Therefore, these solutions are not efficiently designed concerning the optimal use of materials. There have been several research projects on composite beams made of steel and timber [8–12], but are mostly missing the demountability or reuse considerations.

Aiming to address the above-mentioned gaps, our research proposes a novel construction system consisting of demountable hybrid steel and cement-free concrete elements (Fig. 2). The main target is not only to ensure sufficient fire resistance, but also to reduce the losses of gray energy and raw material through a modular structure

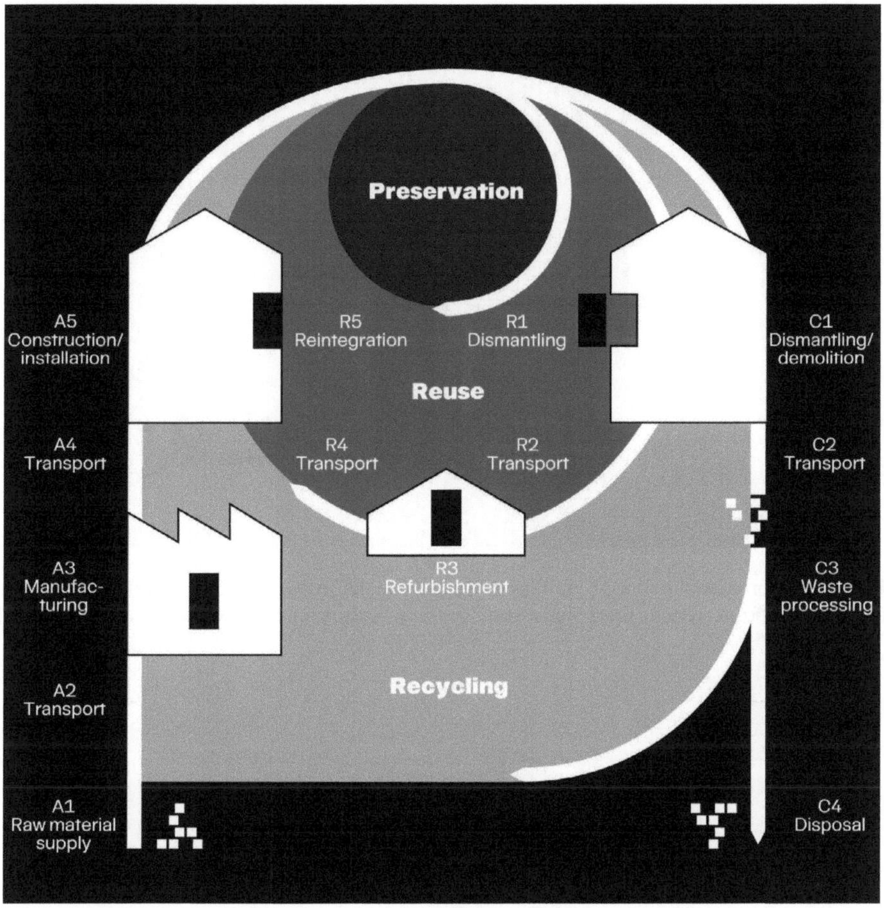

Fig. 1 Stages of circular economy in construction © ZHAW Institut Konstruktives Entwerfen und Baubüro in situ [3]

(as seen in Fig. 3) while enabling the disassembly of buildings, and increasing the overall structural performance of the system. Authors have looked into and examined the fire resistance aspect of this system through experimental and simulation studies and have confirmed the compatibility of the proposed system for both 30 and 60 min in a five-story building. A detailed discussion of these studies can be found in the paper by Thalmann et al. [13].

This research paper specifically focuses on the development of joints and connections within the proposed hybrid system, which facilitates the assembly and disassembly process while ensuring structural integrity, performance, and circularity.

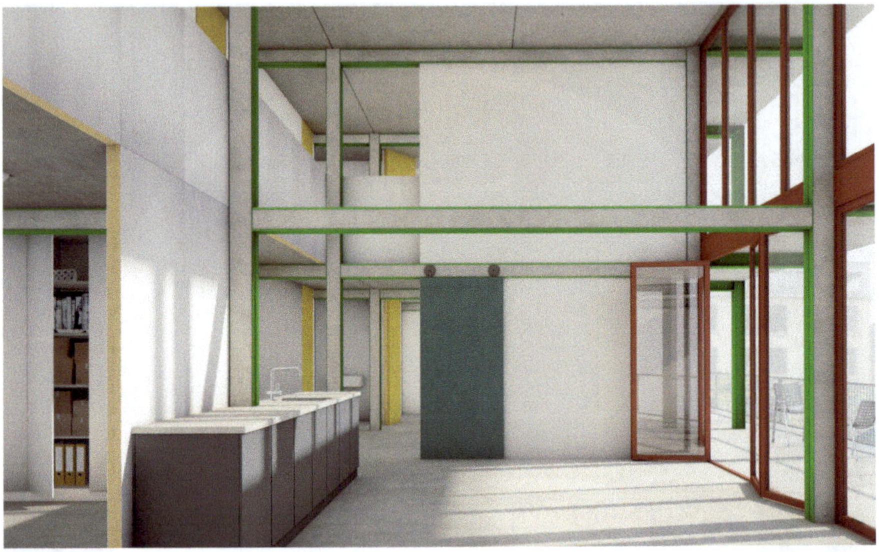

Fig. 2 Interior view of a concept building designed with the proposed folded steel and cement-free concrete composite system, Case Study Steel House by Ressegatti Thalmann and Mario Rinke [14]

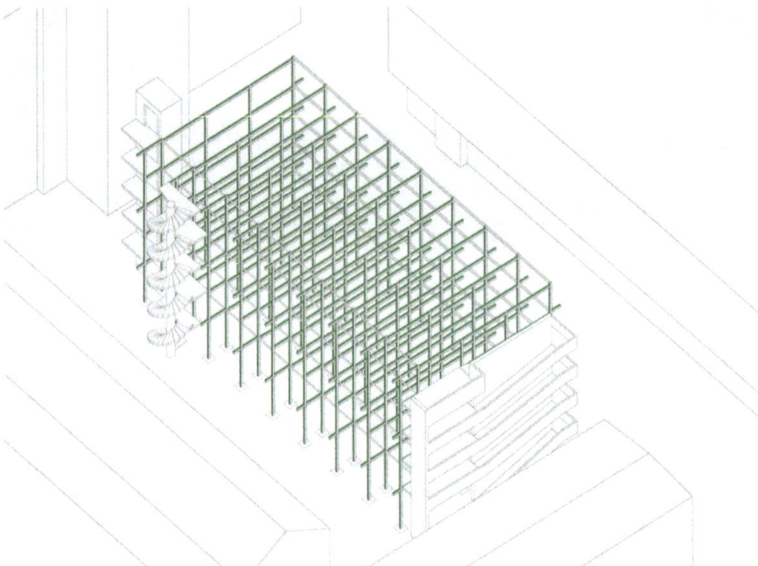

Fig. 3 The general structure of a multi-story building designed with the proposed system, Case Study Steel House by Ressegatti Thalmann and Mario Rinke [14]

2 Structural System

The overall structural system is based on a small-scale steel frame composed of beams and columns. The column's span varies between 3.50 and 5.20 m, and the frame spacing between 2.60 and 4.70 m. These dimensions lie in the common range of application of residential buildings, which is the desired use case for the proposed system. Profile dimensions are designed based on the available Swiss Standards for steel construction [15, 16] and verified through experimental structural tests in the lab of Zürich University of Applied Science (ZHAW) Winterthur [17].

The proposed system consists of two mirrored thin and folded steel C-profiles (S355J2 + N), filled with cement-free clay-based concrete material [17], patented under the name Cleancrete, helping to fireproof the steel cross sections. Cleancrete consists of clay soil or excavated material, aggregates, water, and Oxacrete [18], a natural additive. The material's CO_2 footprint is reduced by 90% in comparison to concrete. In the event of a fire, Cleancrete filling acts as an insulator to protect the inner web and UNP of the columns, ensuring structural reliability for up to 60 min [13]. The steel C-profiles provide bending stiffness and resistance which can be further complemented by the overlaying ceiling slabs connected through dedicated joints. The general construction system is shown in Fig. 4. In this system, a range of connection plates and sheets are developed to enable the connection between standardized modular columns/beams. Following the same principle, connections can be made to adjacent components such as ceilings and facades with a protruding connecting plate, which makes the system expandable as required.

3 Joint Design and Detailing

3.1 Connections Within the Elements

Beam

The beam element is assembled from two thin and folded mirrored C-sections and an inlying web, connected by pre-stressed M16 steel-construction bolts. The bolts are located on the centroid of the section with a longitudinal spacing of 450 mm, derived from experimental 4-point bending tests. In proximity to the ends of the beam, the spacing is adapted to enable the proper connection between beams and columns. A bolted connection is chosen to enable and facilitate the assembly in manufacturing, as well as the disassembly for circularity. Given the foreseen connection with the precast concrete slabs, the inner plate is designed with an increased height, as can be seen in Fig. 5.

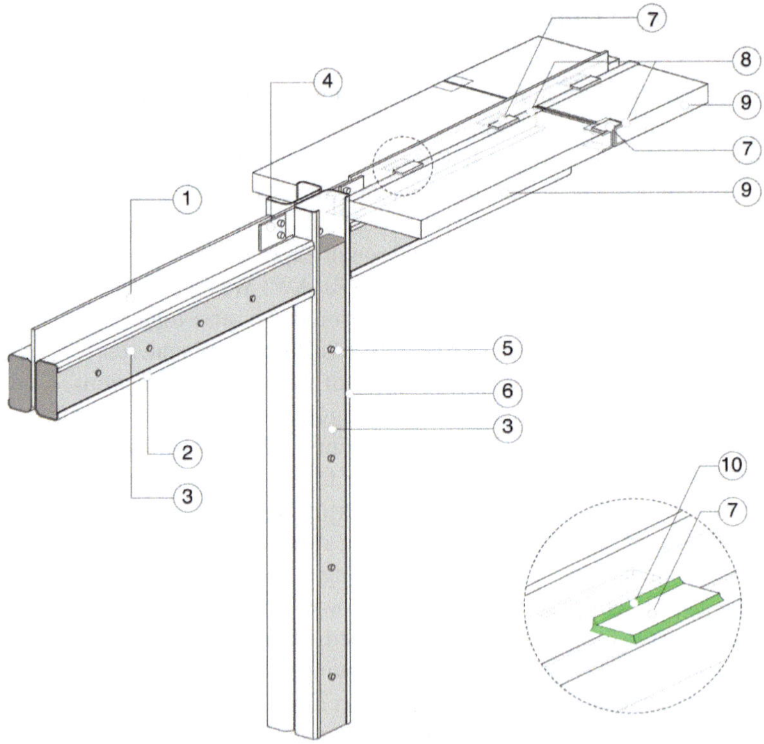

Fig. 4 Axonometric detail of the beam-column connection (1. Inner web 280 * 8 mm, 2. C-profile (beam) cold-formed steel 4 mm, 3. Cleancrete, 4. Fin plate connection, 5. M16 bolt, 6. C-profile (column) cold-formed steel 4 mm, 7. Shear connector plate 120 * 47 * 6 mm, 8. UPE 80, 9. Precast concrete slab h = 80/100 mm, 10. Fillet weld a = 5 mm)

Column

The columns of the system follow the same design principle as the beams. There are two types of columns (illustrated in Fig. 6), mainly due to fire-design considerations. The differentiation is made to accommodate different load-bearing requirements while minimizing material use. The smaller cross-section is akin to the beam design, with a difference in bolt spacing, being 410 mm. The larger cross-section contains an inlaying UNP100 profile, mainly to fulfill the fire requirement for up to 60 minutes, and additionally improve the buckling resistance.

3.2 Connections Between the Elements

Slab to Beam

The pre-cast concrete slabs are manufactured with UPE-Profiles along the axis to enable a welded shear connection, as outlined in Fig. 7. This is achieved with regularly

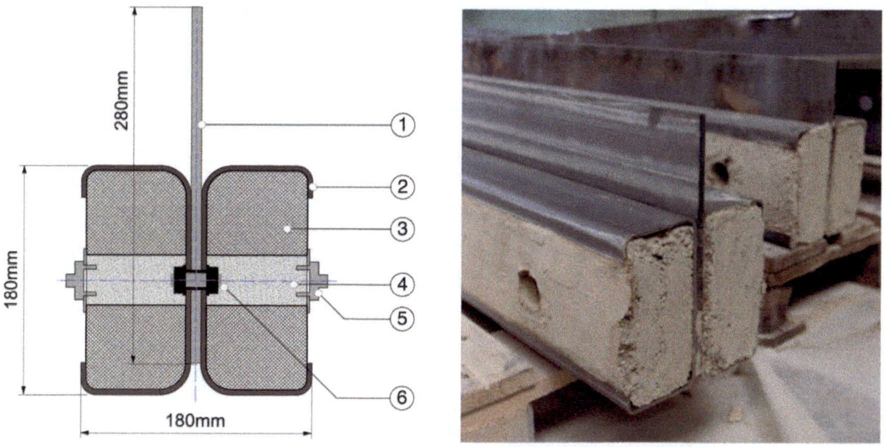

Fig. 5 Beam's cross-section 180 * 180 mm design (left) and real-scale physical prototype (right) (1. Inner web 280 * 8 mm, 2. C-profile cold-formed steel 4 mm, 3. Cleancrete, 4. Mineral wool, 5. Cover, 6. M16 bolt)

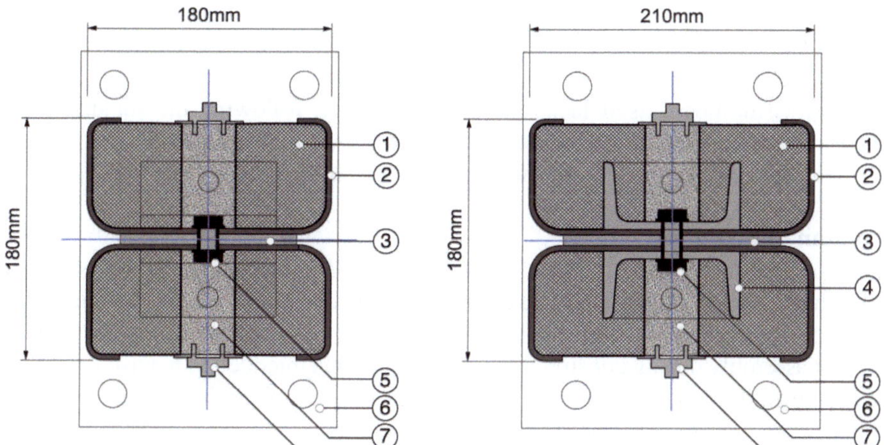

Fig. 6 Column's cross-section 180 * 180 mm (left) and 210 * 180 mm (right) (1. Cleancrete, 2. C-profile cold-formed steel 4 mm, 3. Inner plate 280 * 8 mm, 4. UNP 100, 5. M16 bolt, 6. Head plate FL 280 * 190 * 10 mm, 7. Mineral wool, 8. Cover)

arranged steel shear connector plates, welded to the beam and the UPE profiles. While not perfectly circular, the connection still enables easy disassembly by cutting the small connector plates. It is a product of a lengthy development, with an originally proposed composite floor system using V-shaped recesses in the web of the beam, to accommodate bolted connections to the concrete slab. However, due to production and labor costs, the current connection type was chosen. The composite action is

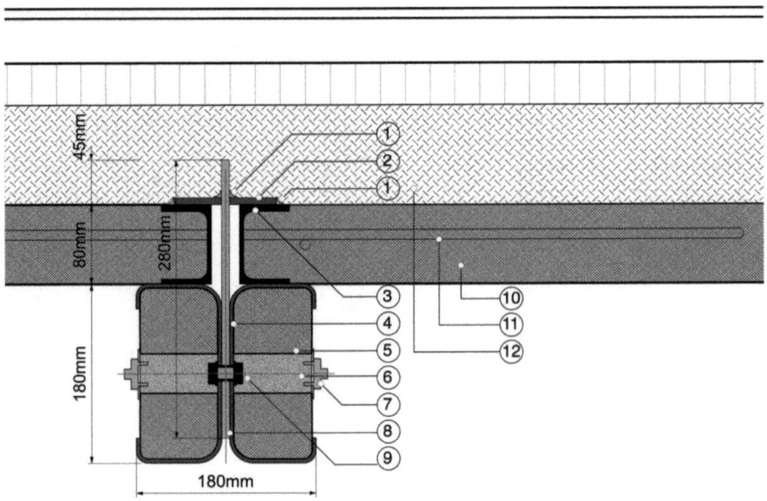

Fig. 7 Joint design between the beam and slab (1. Fillet weld a = 5 mm, 2. Shear connector plate 120 * 47 * 6 mm, 3 UPE 80, 4. C-profile cold-formed steel 4 mm, 5. Cleancrete, 6. Mineral wool, 7. Cover, 8. Inner web 280 * 8 mm, 9. M16 bolt, 10. Precast concrete slab h = 80 mm, 11. Steel reinforcing bar, 12. Limestone chippings)

desired, as the beam itself lacks the needed rigidity, following the standard. The shear-resistant connection with the ceiling slab allows for a slender design for both elements and emphasizes the resource-efficient design philosophy of the system.

Column to Beam

The connection between the beam and column is designed as a slightly eccentric articulated joint (Fig. 8). The web of the columns takes the form of a capital T, on which the flanges of the beam can be connected via two M12 steel bolts. These allow for fast assembly during construction while the joint is intended to behave as a hinge to minimize the propagation of the bending moments through the system.

Column to Column

Each column element is designed to span over two stories, up to 6 m. The connection between two such columns, shown in Fig. 9, is ensured by a head plate butt joint. The headplates are fixed to the C-Profiles, with a slight difference between the two column types. For type 180 * 180 the connection is ensured by a steel angle (LNP-Profile) added on top of the outermost bolts and connected to the head plate with M12 countersunk bolts. For type 210 * 180-U the headplate is welded to the UNP-profiles in the same manner.

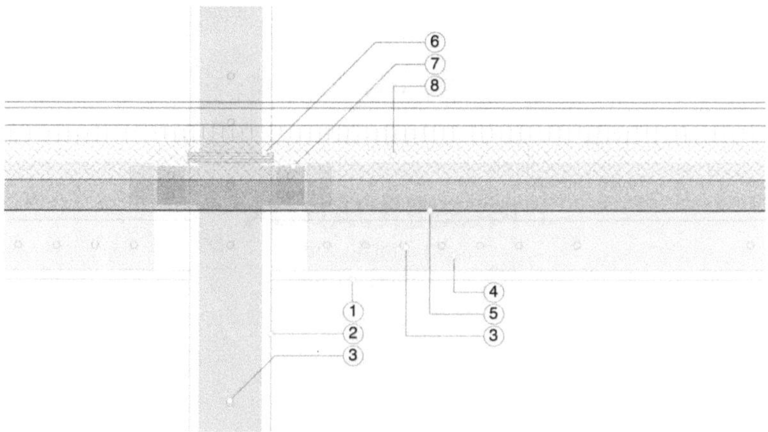

Fig. 8 Joint design between the column and beam (1. Beam 180 * 180 mm, 2. Column 210 * 180 mm, 3. M16 bolt, 4. Cleancrete, 5. Precast concrete slab h = 80 mm, 6. Head plate FL 280 * 190 * 10 mm, Fin plate connection, 8. Limestone chippings)

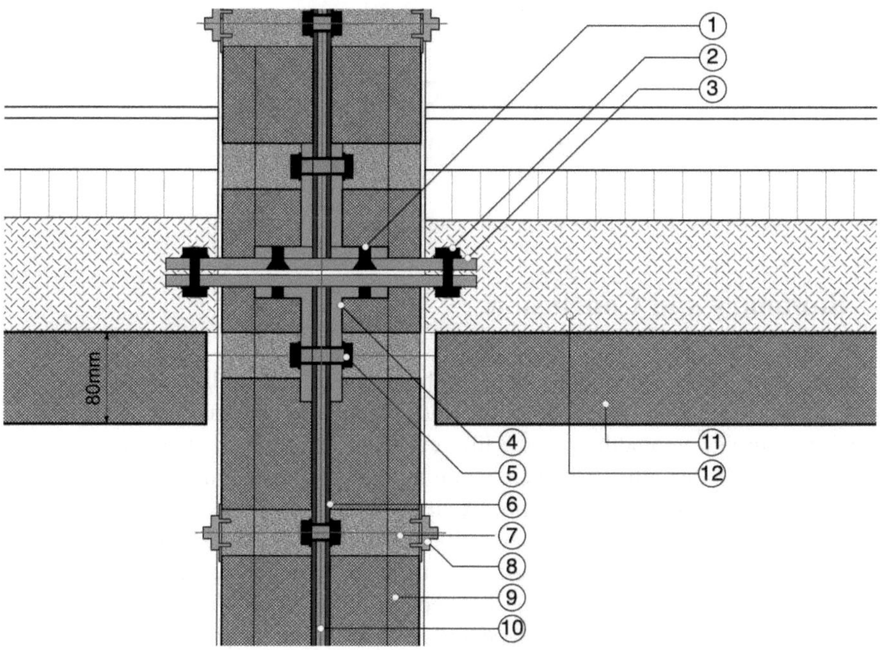

Fig. 9 Joint design column to column for type 180 * 180 (1. M12 countersunk bolt, 2. M12 bolt, 3. Headplate FL 280 * 190 * 10 mm, 4. LNP 100-10, 5. M16 bolt, 6. C-profile cold formed steel 4 mm, 7. Mineral wool, 8. Cover, 9. Cleancrete, 10. Inner web 130 * 8 mm)

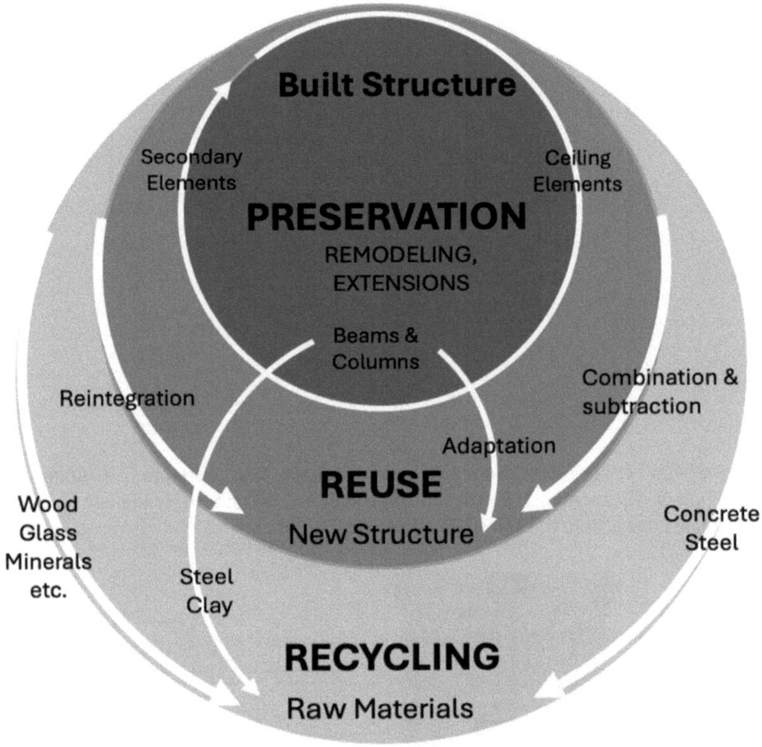

Fig. 10 Stages of circularity in the proposed system

4 Summary and Conclusion

This paper presents the design for disassembly connection detailing the newly developed folded steel and cement-free concrete composite construction system. This novel structure system addresses the fire resistance concerns of steel structures and the circularity in the entire life cycle while reducing the investments in gray energy and raw material, through a flexibly expandable structure due to their connection design. Circularity concerns heavily influence the design of the proposed system. The connection details within these structures are categorized into those within the elements (beams, columns) and between the elements (beam to slab, beam to column, column to column), and the details of each joint design are discussed thoroughly concerning design for disassembly and in alignment with principles of circular economy to facilitate preservation, reuse, and recycle. As summarized in Fig. 10 in the preservation stage the proposed system supports the remodeling, horizontal and vertical expansion of existing structures in various element types (beam and columns, slab elements, and secondary elements such as non-load bearing walls, facades, etc.) through their modular design and connection detail. In the reuse stage, the connection design allows for easy adaptation of components (resizing the columns, beams, and extension or reduction of slab dimensions). Finally, as the crosslink between different materials is held to a minimum, in the recycling phase, the separation of materials is simplified for efficient material recovery. For instance, Cleancrete is easily separable from the steel profiles, facilitating steel and clay recycling.

To summarize, the presented hybrid system and its design of joints is optimized in terms of circularity and fosters a sustainable construction technology that due to its design, leads to an increased robustness over the life cycle of the system.

References

1. Federal Office for the Environment (FOEN) Homepage, https://www.bafu.admin.ch/bafu/en/home/topics/economy-consumption/info-specialists/circular-economy.html#-30093897. last accessed 2024/05/28.
2. European Union Homepage, https://climate.ec.europa.eu/eu-action/climate-strategies-targets/2050-long-term-strategy_en. last accessed 2024/02/08.
3. Stricker, E., Brandi, G., Sonderegger, A., Angst, M., Buser, B., Massmünster, M.: Reuse in Construction, A Compendium of Circular Architecture. p. 10. Park Books, Zurich (2022).
4. Swiss Society of Engineers and Architects: SIA 264:2014 Composite Steel and Concrete Structures. Swiss Society of Engineers and Architects, Zurich (2014).
5. Swiss Society of Engineers and Architects: SIA 264/1:2014 Composite Steel and Concrete Structures - Supplementary specifications. Swiss Society of Engineers and Architects, Zurich (2014).
6. Kodaira, A., Fujinaka, H., Ohashi, H., & Nishimura, T.: Fire resistance of composite beams composed of rolled steel profile concreted between flanges. Fire Science and Technology, 192–206 (2004).
7. Piloto, P. A., Gavilán, A. B., Zipponi, M., Marini, A., Mesquita, L. M., & Plizzari, G. (2013). Experimental investigation of the fire resistance of partially encased beams. Journal of Constructional Steel Research, 80, 121–137.
8. Winter, W., Tavoussi, K., Pixner, T., & Parada, F. R.: Timber-steel-hybrid beams for multi-storey buildings. In: Proceedings of the World Conference on Timber Engineering, New Zealand (2012).
9. Okutu, K. A., Tingley, D. D., Davison, J. B., & Carr, J.: Steel-timber hybrid floors-lowering the embodied impacts of steel frame multi-storey construction. In: Proceedings of the 7th European Conference on Steel and Composite Structures, Italy (2014).
10. Hassanieh, A., Valipour, H. R., & Bradford, M. A.: Experimental and analytical behaviour of steel-timber composite connections. Construction and Building Materials, 63–75 (2016).
11. Loss, C., Piazza, M., & Zandonini, R.: Connections for steel–timber hybrid prefabricated buildings. Part I: Experimental tests. Construction and Building Materials, 781–795 (2016).
12. Loss, C., & Davison, B.: Innovative composite steel-timber floors with prefabricated modular components. Engineering Structures, 695–713 (2017).
13. Thalmann, J.R., Yavaribajestani, Y., Stocker, C., Vigneri, V., Taras, A., Fischli-Boson, P.: Hybridsysteme aus gekantetem Stahlblech und zementfreiem Beton [Sustainable hybrid system made of folded sheet steel and cement-free concrete]. Stahlbau **93**, 782–793 (2024). https://doi.org/10.1002/stab.202400054
14. Reimer T.: Meccano aus Blech und Beton. Werk, bauen+wohnen (9), 27-29 (2017).
15. Swiss Society of Engineers and Architects: SIA 263:2013 Steel Construction. Swiss Society of Engineers and Architects, Zurich (2014).
16. Swiss Society of Engineers and Architects: SN:EN 1993-1-3:2006 Eurocode 3: Design of steel structures - Part 1-3: General rules - Supplementary rules for cold-formed members and sheeting. Swiss Society of Engineers and Architects, Zurich (2006).

17. Thalmann, J., Yavaribajestani, Y., Stocker, C., Kobel, P., Garcia, S., Vigneri, V., Taras, A., Fischli-Boson, P.: Experimental study of resource-efficient folded steel and cement-free concrete composite systems. In: Structures and Architecture: REstructure REmaterialize REthink REuse, Proceedings of the 6th International Conference on Structures and Architecture, Antwerp, Belgium, pp. 255–262 (2025). https://doi.org/10.1201/9781003658641-31
18. Oxacrete Care Homepage, https://oxara.earth/products/oxacrete-care, last accessed 2024/04/21.

Towards Circular Economy in the Fashion Industry: A Regional Case Study in Ticino

Alessandro Fontana, Marzio Sorlini, Andrea Barni, Claudio Capuzzimati, Rosario Pirrotta, and Emanuele Carpanzano

Abstract The transition to a circular economy (CE) in the fashion industry is of high relevance for sustainability. This paper presents a methodological proposal for evaluating the maturity of CE adoption within manufacturing companies. With a focus on the Canton Ticino in Switzerland, where the fashion industry is highly specialized, this study examines the evolving landscape of the sector, highlighting shifts towards integrated logistics and outsourcing. A survey-based approach, complemented by targeted interviews, assesses the maturity of CE adoption among companies. Findings reveal varying levels of knowledge and adoption of CE principles, with significant opportunities for improvement identified. Key enabling factors include technological advancements, efficient reverse logistics, education, policy support, and consumer engagement. Challenges include the need for centralization in garment collection, enhanced sustainability education, and effective policy implementation. The proposed methodology offers insights into CE maturity levels, guiding stakeholders towards sustainable practices and fostering sector-wide circularity. Extension of this methodology to other sectors holds promise for comprehensive assessments and targeted actions, driving sustainable development on a broader scale.

Keywords Circular economy · Maturity assessment · Fashion industry

1 Introduction

The Canton Ticino exhibits in the textile and clothing industry the highest degree of specialization amongst the other sectors, and it also has the highest productivity overall.

According to the studies conducted by the institute of economic research (IRE) [1], the research institute BAK Basel in 2014 [2] USTAT in 2015 [3] and, there is a clear shift in the fashion landscape in Ticino, evolving from a purely manufacturing sector

A. Fontana (✉) · M. Sorlini · A. Barni · C. Capuzzimati · R. Pirrotta · E. Carpanzano
University of Applied Sciences and Arts of Southern Switzerland, Lugano-Viganello, Switzerland
e-mail: alessandro.fontana@supsi.ch

A. Podleisek et al. (eds.), *Advances in Circular Economy: Bridging Research and Practice*, Springer Proceedings in Earth and Environmental Sciences,
https://doi.org/10.1007/978-3-032-21547-5_10

to a true meta-sector. The term defines the transformation of integrating, in addition to traditional manufacturing activities, services such as consulting, logistics and other ancillary functions that support the primary industry. Most of the newly created businesses are now active in integrated logistics dedicated to fashion. Companies are integrating activities beyond the traditional textile and clothing scope into the value chain, embracing a broader and more innovative definition of the meta-sector [4]. Despite this evolution, design is a core activity for Ticino fashion industries that remains locally managed, allowing companies to directly influence and guide the sustainability and circularity of products. Indeed, many are the initiatives that are currently put in place to reduce the use of resources and the environmental burdens generated, but a common vision and a structured way of intervening is often lacking.

This study is developed in collaboration with Ticinomoda Association, which includes both companies directly involved in product manufacturing and entities from ancillary sectors. In this context, the paper aims to present the development of a circularity assessment framework for evaluating the maturity of circular economy adoption within fashion industries in Ticino. It contributes to deepening the understanding of circular economy adoption maturity among manufacturing companies, shedding light on the associated challenges and opportunities. The framework is constituted by a survey and interviews meant to investigate the status of CE implementation within fashion companies, to highlight current and future trends to be adopted, barriers and opportunities. The paper begins with a state-of-the-art review highlighting the weaknesses of the linear economy and the need for a shift towards a circular economy that must be guided by quantitative assessments and tools for companies CE maturity evaluation. Building upon these findings, Section 3 presents the methodology developed for the maturity analysis that brought to the survey design, while Sect. 4 reports the test and validation of the methodology that has been performed in a qualitative way proposing surveys and interviews to a selected group of Ticinomoda members. Eventually, Section 5 discusses the lessons learned from the methodology application, the possible evolution of the methodology and its extension to other industrial sectors.

2 State of the Art

Much of today's fashion industry follows a linear economy model that does not consider concepts such as extending product lifespan, reusing waste, or managing end-of-life responsibly but is based on the belief that resources are infinite [5, 6]. While this awareness is growing, there is still much to be done in addressing these issues effectively. The linear economy, with its traditional take-make-waste approach leading to excessive consumption and frequent disposals, resulting in waste and impacts, is only marginally affected by concerns expressed by some industry stakeholders, while a transition to circular practices could offer a virtuous response to contemporary challenges [7].

According to the article "Vision of a circular economy for fashion" published by the Ellen MacArthur Foundation [8], circular economy (CE) is based on three principles, all guided by design: ensuring products are used more and for longer, designed to be reused once they reach the end of their life, and using materials with reduced environmental impact, i.e., recycled or from renewable sources. Implementing circular economy principles for fashion allows for better products and services for customers, contributes to resilient supply chains, and, of course, reduces environmental impact [9]. The circular economy can also generate new competitive advantages for companies: it is indeed possible to create new growth opportunities that are distributed, diversified, and inclusive.

While sustainability goals in the fashion industry encompass broader environmental, social, and economic aspects, circular economy (CE) focuses on closed-loop systems, emphasizing reducing, reusing, and recycling to minimize waste and resource use [10]. Although CE initiatives are growing, they remain sporadic and less formalized compared to sustainability efforts [11]. The fashion industry's inclination towards sustainability often overlooks CE's specific emphasis on resource management and innovative business models [12].

Analysing the available literature on the topic of CE applied to the Fashion sector, several macro business models can be identified [13]: (1) repairing focuses on refurbishing existing products to extend their lifespan and reduce resource usage, exemplified by Patagonia's repair services [14]. (2) Remaking involves reconfiguring existing products for new purposes through disassembly and reuse of components, as seen in brands like Freitag's use of reclaimed materials [15]. (3) The use of bio-based and recycled materials substitutes non-renewable resources, with brands like Stella McCartney emphasizing resource efficiency [16]. (4) Creation from production waste repurposes scraps, demonstrated by companies like Ecoalf [17]. (5) Take-back programs like H&M's Garment Collection recycle end-of-life products [18]. (6) Second-hand sales, facilitated by platforms like Reflaunt, promote reuse [19]. (7) Product sharing via platforms like Tulerie fosters a sharing economy [20]. (8) Servitization offers rental or pay-per-use services reducing consumption, and this is supported by platform such as Nuuly [21].

Regarding the circularity maturity assessment, research can be supportive in identifying what should be specifically assessed and considered about CE evaluations [22]. In the last decade, many studies attempted to propose specific metrics or indicators for assessing CE at different levels (e.g., company, regional/national), and to develop comprehensive CE assessment methods. Both these streams of studies have been recently reviewed from different angles; among the others, the authors of [23, 24] have focused on CE metrics or indicators, while [25, 26] have targeted CE assessment methods. Despite the intense scholarly debate on CE assessment, a common framework has not been agreed on, as clearly pointed out by Niero and Kalbar [27]. In addition, the studies on CE indicators and assessment methods do not always build upon each other; in other words, new CE assessment methods have been developed while overlooking possibly interesting results achieved in systematic research on CE metrics and indicators.

Several studies have explored various approaches to assess the level of CE implementation in organizations. Fonseca et al. [28] developed an online questionnaire-based assessment for Portuguese organizations, while similar approaches were followed for Spanish, Basque, and Italian Small and Medium Enterprises (SMEs) [29, 30]. However, these national evaluations may not be suitable for organization-based assessments on a large scale. While they aim to gather general data and identify barriers to CE implementation, they do not guide companies towards higher CE performance levels. Other studies have focused on sector-specific or niche-based assessment methods, such as those developed by Berzi et al. [31] and Bilal et al. [32].

Additional evaluation tools have been developed to provide companies with indications of their CE performance and readiness. Examples include Circulytics by the Ellen MacArthur Foundation [33], Circle Assessment by Circle Economy [34], and CE readiness assessment by the Technical University of Denmark [35]. These tools offer user-friendly experiences for respondents but may lack transparency on methodologies and theoretical foundations. While practitioner-oriented CE assessment tools can calculate aggregate scores for CE performance, their scientific reliability remains unclear.

The methodology adopted in this paper builds upon the review carried out by Sacco et al. [36] which highlights the need to develop a tool for CE assessment at firm level and identifies its main requirements. Specifically, the authors presented several categories for the classification of the CE aspects which have been further processed in Sect. 3.

3 Methodology: Survey and Semi-structured Qualitative Interviews

To analyse the Association members' knowledge and practical adoption of CE themes and practices, a survey has been designed and conducted with 9 companies directly involved in the Fashion industry supply chain. As agreed with the Association, auxiliary sector subjects such as logistic and consultancy operators were not involved since the analysis has been focused on the manufacturing context, thus the investigation areas would be different and the obtained results not easily comparable. With the aim of providing a comprehensive overview of the sector's maturity level regarding CE issues, it was necessary to select a unique reference model against which to analyse the performance of all interviewed companies. Literature analysis on maturity models for EC led to selecting the approach proposed by Sacco [36]. The method divides the survey into different categories aiming to cover all relevant aspects for CE. This approach has been partially revised and adapted by adding some questions of particular interest for the reference sector, and revising dimensions and themes suggested by additional consulted scientific articles.

Specifically, the initial list structure was adjusted to align with the fashion sector. A series of studies focused on circular economy in fashion carried out by Boscacci et al.

[37], Colasante et al. [38], Dissanayake and Weerasinghe and the Ellen MacArthur Foundation [39] were analysed to capture various aspects, which were then transformed into questions and assigned to categories according to the taxonomy defined below in this paragraph. This was undertaken because the reference framework developed by Sacco et al. is generally addressing circularity, without considering the specificities of the fashion compartment. Consequently, a gap analysis was performed to identify the shortcomings of the model, which were subsequently addressed using previously cited papers. It was found that although the proposed macro categories were consistent with the fashion sector, the suggested questions and responses were often too high-level, consisting of binary answers (yes or no), or not relevant for the fashion industry. This includes a semi-quantitative analysis, where respondents were asked to indicate the percentage of waste managed by each approach. More details regarding the revisions made to the initial model proposed by Sacco et al. are presented below as the various categories are described. The integrated literature allowed the authors to give greater emphasis to areas related to human resource management, logistics, and technological development by refining the questions and modifying the responses to better align with the fashion industry. The survey consists of a total of 44 questions divided into 5 categories, which corresponds to an average completion time of 25 minutes, in addition to an initial section meant to profile the interviewed company and respondent, thus collecting information on the company dimension the main activities carried out and the reference market type. The five categories are described as follows. The first section *Communication and Circular Business Models* aims to evaluate the company's infrastructure holistically, encompassing its physical, relational, and immaterial aspects related to EC. In particular, the key themes explored include participation in EC-related associations, internal communication regarding circular economy practices, analysis of the applicability of circular business models as well as the revenue generated, and examination of waste management practices through waste destination analysis. For instance, in this case, the initial framework included binary questions to determine whether a company was active in this area. This was revised to incorporate a more detailed understanding of the specific approaches suggested by the literature. The second section *Suppliers and Logistics* explores the evaluation criteria companies impose on suppliers concerning EC principles and how these criteria impact supplier qualification and requalification. Additionally, it assesses reverse logistics towards customers and the subsequent management of returned products. Compared to the reference framework, more detailed elements were added regarding aspects of reverse logistics. Instead of simply determining whether it was implemented, the investigation included the use of potential KPIs and, importantly, the methods of managing products once they are returned. The third section *Marketing and Sales* aims to evaluate the actions that the interviewed companies undertake in the field of marketing and sales to ensure a commercial return on their EC initiatives. The novelties introduced by the authors involve requesting more information about the methods of sharing knowledge related to the circular economy within the companies. Specifically, detailed elements were added to better understand not only the maturity of communication but also the primary channels used for dissemination. The fourth section *Technological*

Development and Circular Economy Practices aims to assess the company's investment in technologies, processes, and materials enabling EC. As already considered by Sacco [36], categories include cost reduction, circular design practices, intellectual property protection, innovative materials, and technology development for EC improvement. With respect with the initial model, the structure of the questions remained unchanged. Responses were modified, focusing on technologies suitable for the circular economy in the fashion sector. This included, among other aspects, exploring into blockchain technologies, augmented reality, and big data. The fifth section *Challenges and Opportunities* analyses obstacles and prospects related to the implementation of circular business models. Obstacles (financial, administrative, political, organisational) refer to the potential reasons that make it more difficult for companies to implement new concepts, practices, and business models. On the other hand, prospects (acquiring new customers, reducing impacts, extending lifecycle) define all the potential advantages that a company could gain from implementation. Similarly to the previous category, specific details relevant to the fashion sector were also added in this case. These pertain to design challenges for creating new products, the limited availability of raw materials, and the lack of standards for recovered products. In terms of opportunities, the previously cited literature studies helped identify more prospects for companies, such as job creation and income sources, gaining competitive advantage through cost reduction, and waste generation. To ensure the effectiveness and relevance of the survey questions to the sector and verify the time needed to fulfil the job, the questionnaire has been submitted as a test to two companies belonging to Ticinomoda that are frequently collaborating with the authors. During the development of the survey, potential areas for further exploration were identified. These elements were addressed at a later stage to prevent the survey from becoming too lengthy. So, to complete the maturity assessment, interviews have been designed and carried out with a selected group of companies. These interviews served as a post-survey follow-up, allowing for a deeper exploration of the quantitative aspects addressed in the survey sections. Specifically, each company's survey responses were examined to tailor discussions on the most relevant topics. The interview frameworks were constructed considering the survey responses and the priority areas identified during the initial phase. The interview format can be classified as semi-structured, as key questions and topics have been identified, but we maintained the flexibility to deviate to explore additional aspects. Despite the interviews have been tailored to the characteristics of the individual companies, a common structure was defined for each area of focus so that the responses obtained would be comparable and could provide effective and insightful analysis. The focus areas that were especially taken into consideration concerned challenges and opportunities associated with implementing, managing, and sustaining circular business models. More specifically, the implementation of business models was examined, focusing on how each model is managed within the companies and the resulting human and financial implications. In the area of waste management, after identifying management techniques through the survey, a detailed analysis was conducted on the disposal and recovery of certain products and materials, discussing potential issues and how best practices and circular economy approaches can support this process. Lastly, the survey highlighted the opportunities and challenges that companies recognize in circular business models. During the interviews a detailed discussion followed on how joint or ad hoc projects can maximize benefits and overcome the barriers to the implementation and transition towards circular economy.

4 Results: Test and Validation of the Methodology

The circularity maturity methodology has been tested and validated submitting the survey to a limited number of Ticinomoda members, specifically nine companies plus two firms that has been used as preliminary testers. Moreover, out of the eleven companies investigated with the questionnaire, six expressed their willingness to respond also to the interviews. Since the restricted number of companies investigated, the analysis of the results cannot be statistically relevant, but it still allows a verification of the designed approach and a check on the possibility to extract meaningful information and data from the structured interaction with the companies.

Concerning the survey, the section about the company profile shows that respondents are evenly divided between large companies (4 companies) and SMEs (5 companies), offering a comprehensive perspective on company size. Additionally, various roles within the fashion supply chain are represented: the results indicate that respondents can occupy multiple roles, with 8 engaged in design activities, 7 in production, one person in third-party processing, 6 people directly marketing their products, and one person involved in logistics operations. The self-assessed competency on the topics shows that 2 respondents claim to have a low knowledge of the main CE concepts, 3 an average knowledge, and 4 an advanced one. The analysis of the results and the conducted interviews is then meant to validate if the values indicated at the beginning of the survey by the respondents reflect the real situation. For each of the above cited category of the survey, the main results are illustrated in the following.

Communication and Circular Business Models. A portion of respondents (2) are affiliated with organizations involved in the circular economy. Similarly, 2 respondents internally communicate Circular Economy (EC) practices and principles. Among the various circular business models, those focusing on the utilization of bio-based/recycled materials and the conversion of production waste into products are most prevalent; companies affirm an increasing trend toward the use of bio-based and recycled materials. In fact, it was found that all the companies surveyed are making efforts to reduce the use of virgin materials, both in the making of the garment and in its preparation for sale. This is thanks to an increase in the purchase of materials certified as recycled from their suppliers. Anticipated developments in the near future include the implementation of end-of-life product take-back and reuse, alongside an increase in second-hand sales; according to some interviewees, second-hand sale is, to date, mostly aimed at high-end brands and most often is not carried out directly by the brand but by customers who decide to resell an old garment or accessory to a new user through multi-sided platforms managed by third parties. The reasons for the limited presence of brands in these initiatives are mainly due to the difficulty of reaching the critical mass necessary to justify the investments that would be required. Conversely, models perceived as less applicable encompass sharing products among customers via platforms and servitization; the survey and particularly the interviews revealed that the possibility of establishing such services for a specific brand is low. This is because the management of such a business plan would create the need for product administration even in its use phase and not only in the production and sales phase (Fig. 1).

On average, circular business activities contribute 9% to respondents' revenues. This figure ranges from 0% to 15% for 7 respondents, and between 15% and 30% for the remaining 2. Notably, over half of the waste generated is earmarked for recycling or reuse. Conversely, energy recovery ranks as the least utilized waste management method, accounting for only 11%. Furthermore, less than 30% of waste stemming from manufacturing activities is ultimately disposed of in landfills.

Suppliers and Logistics. The results of this section aim to understand the evaluation of suppliers from a circular perspective and the applied methods of reverse logistics for product collection at their end of life. Regarding the first aspect, 5 respondents apply CE principles in assessing suppliers, with over 50% of suppliers partly adhering to these principles. 4 companies have considered replacing non-compliant partners, while the remaining 5 companies prefer to create a partnership and invest directly to help suppliers in enhance their circular practices. Suppliers are assessed via audits or questionnaires, leading to the definition of suppliers' profiles ranked via their sustainability and circularity performances. Regarding reverse logistics activities, the

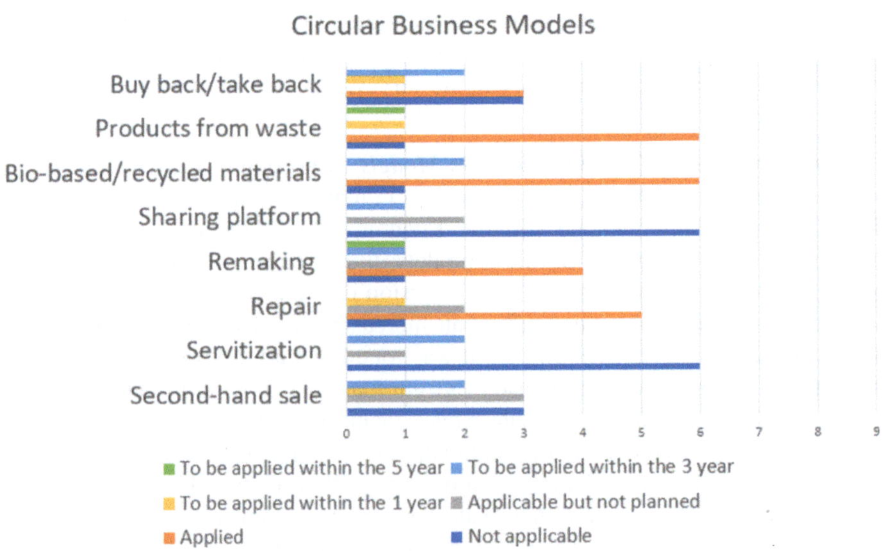

Fig. 1 Circular business models

analysis revealed that among the interviewed companies, recycling (3 respondents), reuse (2 respondents), and landfill disposal (2 respondents) are the management methods applied.

Marketing and Sales. Among the sample, 6 respondents do not communicate information related to the circular economy through communication channels. Conversely, the remaining 3 utilize a variety of channels, with sustainability reports, social networks, and websites being the most prominent choices.

Technological Development. This section discusses the importance of understanding the level of investment companies make in technologies, processes, and materials that support the Circular Economy (CE). Key areas of examination include the implementation of CE approaches and technologies, circular design practices, the types of materials used, and the enabling technologies involved. The utilization of recovered raw materials for new product manufacturing stands out as the most prevalent approach within the realm of Circular Economy (EC). Looking ahead, there is an anticipated surge in the adoption of strategies centred around Product Circular Design and Dematerialization. Conversely, methodologies such as Asset Sharing and Industrial Symbiosis appear less feasible for widespread implementation. A noticeable shift in CE practices involves an increasing inclination towards the adoption of shared, recycled, or reused packaging solutions (Fig. 2.).

Furthermore, there is a discernible preference for certified materials of origin, often chosen for their lower environmental impact compared to substitutes. However, these materials are typically sourced from third-party suppliers rather than internal processes. The exploration of bio-based materials warrants attention to ascertain

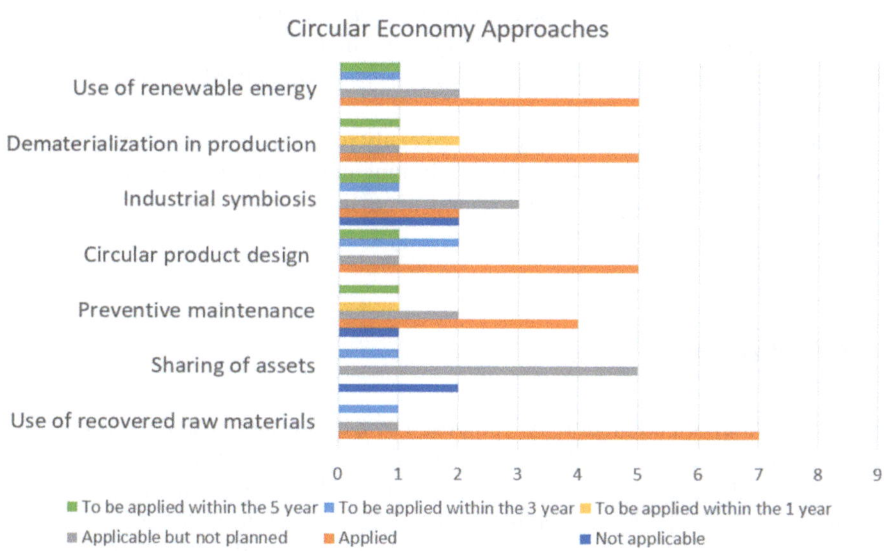

Fig. 2 Circular Economy Approaches

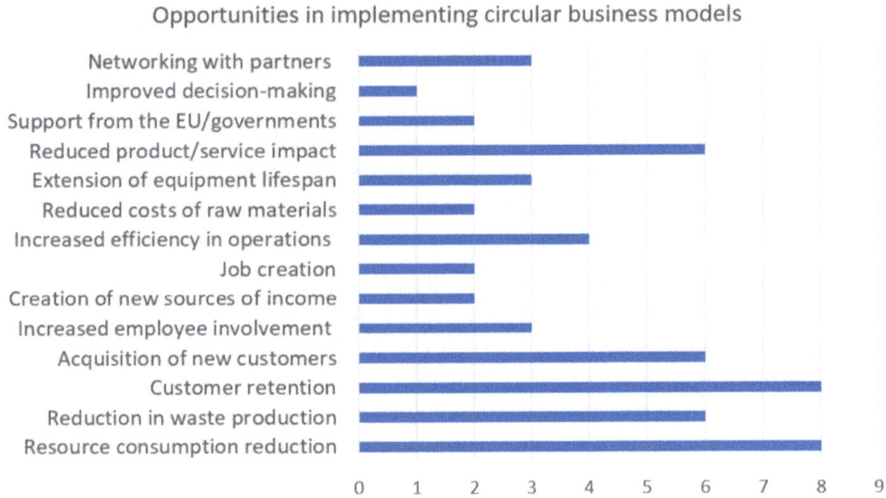

Fig. 3 Opportunities in implementing circular business models

whether they are integral to forward-thinking corporate strategies or already in active use, particularly among sectors like cotton or wool production. Among emerging technologies, Big Data and Robotic Systems are garnering significant consideration. Despite these advancements, there remains a general deficiency in the widespread application of enabling technologies for EC initiatives.

Challenges and Opportunities. The primary challenges hindering the implementation of CE initiatives include a scarcity of partners and inadequate material availability, identified by 6 respondents. Additionally, ineffective recycling policies and the complexity of management and planning processes pose significant barriers, cited by 5 respondents.

In terms of opportunities, companies are keenly focused on resource consumption reduction (8 respondents), waste and environmental impact reduction (6 respondents), as well as customer retention and acquisition, each noted by 8 and 6of respondents, respectively (Fig. 3).

5 Conclusions

This study presents a methodology, composed by a questionnaire and interviews, for the evaluation of Circularity Maturity of fashion industries. While there has been progress in embracing CE principles, significant challenges remain and a structured vision on how to apply those principles and translate them into reality is still lacking. The analysis highlighted the implications of implementing the CE macro business model and identified the pros and cons of various tools for assessing circularity, revealing gaps to be filled. The developed model addressed these gaps and offered a qualitative view of the fashion sector's circularity landscape in Ticino Canton Ticino's textile and clothing industry in Switzerland thanks to a test performed on a selected group of companies. Interviews with companies underscored their commitment to exploring the opportunities presented by the identified business models.

The performed test allowed to verify that the survey is clear and easy to be completed, needing less than 25 minutes estimated, even when the respondent is not an expert on the circularity and sustainability field. The results are not providing conflicting information and, more interesting, can be easily translated in quantitative data. This is shown by the obtained results analysis despite the number of the respondents is too low to extract some data of statistical relevance. It is thus expected that submitting the survey and the interviews to a larger set of companies, also outside the Ticino region and Switzerland, could derive interesting and meaningful sectorial trends. If the questionnaire allows the collection of data that can be translated in quantitative evaluations of the circularity level and on the gaps to be filled to improve the circularity performances, the presence of structured interviews allows to deepen some specificities of the single companies that can provide a more detailed idea of the practices application, but that can also represent benchmarks to be exploited as a reference for the whole compartment. In addition to that, the methodology is not only relevant for statistical usage, but offer also to the single companies a vision on the circularity maturity and a way to identify possible improvement path to better guide and focus the efforts. In this context, the survey results can be exploited to plan future research and innovation activities to be carried out in a collaboration amongst companies and research institutes. Even with few responses, the questionnaire indeed allowed to identify some priority circularity themes to be addressed by the Ticino fashion firms, such as: *Design for circularity*, *End of life management* (of products and wastes), *Supply chain traceability and certification.* Those common theme of interest have been considered by Ticinomoda to stimulate the activation of research and development projects meant to develop real solutions to support the transition towards CE.

Moreover, the methodology employed in this study provides a solid foundation that can be adapted for other industrial sectors. This creates an opportunity for in-depth assessments and strategic actions to promote sustainable and circularity practices. The developed methodology holds significant potential for expanding its application to other industrial sectors, allowing for the creation of sector-specific studies on circular economy maturity. This can be achieved by tailoring the survey to suit the characteristics of each respective sector. For instance, applying the first section of the survey (communication and circular business models) to the automotive sector, aspects such as handling of the end-of-life phase of vehicles and parts can be integrated [40]. Additionally, strategies for reducing waste during the manufacturing phase and examining how companies implement efficient resource management and waste reduction practices can be considered. By customizing the survey questions and analysis to address the unique challenges and opportunities within different industries, a comprehensive understanding of circular economy practices and their maturity levels can be obtained. Moreover, extending the methodology to various sectors enables a broader assessment of the circularity landscape, facilitating benchmarking, identification of best practices, and targeted interventions for enhancing circularity across diverse economic domains. Through this approach, stakeholders in various industries can gain valuable insights into their circular economy performance, driving informed decision-making and fostering sustainable development practices on a broader scale.

Acknowledgments This study has been supported by Ticinomoda. The authors would like to thank to the Association and the members that responded to the survey and the interviews for their valuable contributions and the interest shown on the project.

References

1. R. Maggi and V. Mini, "La catena di valore transfrontaliera: il potenziale dei sistemi integrati di produzione," Istituto di ricerche economiche (IRE): Osservatorio delle politiche economiche (O-Pol), 2013.
2. A. Eberli, S. Hilber, Krummenacher, R. Rufer, A. Wagner and F. Zainhofer, "Analisi dei settori ticinesi: benchmarking internazionale e smart specialisation," 2014.
3. O. Gonzalez, S. Petrillo, E. Stephani and S. Walker, "I COMPARTI ECONOMICI - Struttura ed evoluzione del tessuto economico ticinese," 2015.
4. A. Fontana, D. Corti, A. Barni and F. Moltoni, "Sustainability Certifications and Labels for the Fashion Industry: Selection Guidelines," in *Business Models and ICT Technologies for the Fashion Supply Chain*, 2019.
5. E. Carpanzano, R. Fornasiero, A. Chiodi and L. Carneiro, "Research Issues on Customer-Oriented and Eco-friendly Networks for Healthy Fashionable Goods," in *IFIP Advances in Information and Communication Technology*, 2010.
6. C.-W. Ki, S. M. Chong and J. E. Ha-Brookshire, "How fashion can achieve sustainable development through a circular economy and stakeholder engagement: A systematic literature review," *Corporate Social Responsibility and Environmental Management,* vol. 27, no. 6, pp. 2401–2424, 2020.
7. Europea Environment Agency, 27 February 2023. [Online]. Available: https://www.eea.europa.eu/publications/eu-exports-of-used-textiles.
8. Ellen MacArthur Foundation202, 2020. [Online]. Available: https://www.ellenmacarthurfoundation.org/our-vision-of-a-circular-economy-for-fashion.
9. E. Carpanzano and R. Fornasiero, "Advances in customer-oriented manufacturing and value chain management. International Journal of Computer Integrated Manufacturing," 2017.
10. R. Koppelaar, S. Pamidi, E. Hajósi, L. Herreras, P. Leroy, H.-Y. Jung, A. Concheso, R. Daniel, F. Francisco and C. Parrado, "A Digital Product Passport for Critical Raw Materials Reuse and Recycling," *Sustainability,* 2023.
11. M. Geissdoerfer, P. Savaget, N. M. Bocken and E. J. Hultink, "The Circular Economy—A new sustainability paradigm?," *Journal of Cleaner Production,* vol. 143, pp. 757–768, 2017.
12. M. Colucci and A. Vecchi, "Close the loop: Evidence on the implementation of the circular economy from the Italian fashion industry," *Business Strategy and the Environment,* vol. 30, no. 2, pp. 856–873, 2021.
13. A. D. Marques, A. Marques and F. Ferreira, "Homo Sustentabilis: circular economy and new business models in fashion industry," vol. 2, no. 2, 2020.
14. "Worn Wear is our program to trade in and buy used Patagonia® gear," [Online]. Available: https://wornwear.patagonia.com/.

15. "Il nostro cammino all'insegna dell'economia circolare," [Online]. Available: https://www.freitag.ch/it/circularity-roadmap.
16. "Circularity," [Online]. Available: https://www.stellamccartney.com/it/it/sustainability/circularity-2.html.
17. "Ecoalf Sustainability Report," [Online]. Available: https://ecoalf.com/en/pages/memoria-de-sostenibilidad?mdApp_countryCodeDomain=IT.
18. "Let's close the loop," [Online]. Available: https://www2.hm.com/en_gb/sustainability-at-hm/our-work/close-the-loop.html.
19. "Balenciaga Launches Resale Programme," [Online]. Available: https://www.businessoffashion.com/news/sustainability/balenciaga-launch-resale-kering-circularity-sustainability-second-hand/.
20. "Thinking About Fashion Rentals? Consider These 7 Services," [Online]. Available: https://www.vogue.com/article/clothing-rental-subscription.
21. "Sustainability," [Online]. Available: https://www.nuuly.com/rent/sustainability.
22. L. Canetta, A. Fontana, M. Foletti and M. Tschanen, "Sustainability Assessment of Various Circular Economy Scenarios in the Consumer Electronics Sector," in *IEEE International Conference on Engineering, Technology and Innovation (ICE/ITMC)*, 2018.
23. H. S. Kristensen and M. M. Alberg, "A review of micro level indicators for a circular economy—moving away from the three dimensions of sustainability?," *Journal of Cleaner Production,* vol. 243, 2020.
24. B. Corona, L. Shen, D. Reike, J. R. Carreón and E. Worrell, "Towards sustainable development through the circular economy—A review and critical assessment on current circularity metrics," *Resources, Conservation and Recycling,* vol. 151, 2019.
25. C. Sassanelli, P. Rosa, R. Rocca and S. Terzi, "Circular economy performance assessment methods: A systematic literature review," *Journal of Cleaner Production,* vol. 229, pp. 440–453, 2019.
26. E. R. Lindgreen, R. Salomone and T. Reyes, "A Critical Review of Academic Approaches, Methods and Tools to Assess Circular Economy at the Micro Level," *Sustainability,* vol. 12, 2020.
27. M. Niero and P. P. Kalbar, "Coupling material circularity indicators and life cycle based indicators: A proposal to advance the assessment of circular economy strategies at the product level," *Resources, Conservation and Recycling,* vol. 140, pp. 305–312, 2019.
28. L. M. Fonseca, J. P. Domingues, M. T. Pereira, F. F. Martins and D. Zimon, "Assessment of Circular Economy within Portuguese Organizations," *Sustainability,* 2018.
29. J. Santos, C. Jaca, V. Prieto-Sandoval and M. Ormazabal, "An overview of the circular economy among SMEs in the Basque Country: A multiple case study," *Journal of Industrial Engineering and Management,* 2016.
30. M. Mura, M. Longo and S. Zanni, "Circular economy in Italian SMEs: A multi-method study," *Journal of Cleaner Production,* vol. 245, 2020.
31. L. Berzi, M. Delogu, M. Pierini and F. Romoli, "Evaluation of the end-of-life performance of a hybrid scooter with the application of recyclability and recoverability assessment methods," *Resources, Conservation and Recycling,* vol. 108, pp. 140–155, 2016.
32. A. R. Nasir, K. I. A. Khan, M. Bilal and T. M. Jamaluddin, "Current state and barriers to the circular economy in the building sector: Towards a mitigation framework," *Journal of Cleaner Production,* 2020.
33. Ellen MacArthur Foundation, 2020. [Online]. Available: https://www.ellenmacarthurfoundation.org/resources/circulytics/overview.
34. [Online]. Available: https://www.circle-economy.com/digital-offering.
35. D. C. Pigosso and T. C. McAloone, "Making the transition to a Circular Economy within manufacturing companies: the development and implementation of a self-assessment readiness tool," *Sustainable Production and Consumption,* vol. 28, pp. 346–358, 2021.
36. P. Sacco, C. Vinante, Y. Borgianni and G. Orzes, "Circular Economy at the Firm Level: A New Tool for Assessing Maturity and Circularity," *Sustainability,* 2021.
37. E. Boscacci, "Circular Economy in the Fashion Industry: turning waste into resources," 2018.

38. A. Colasante and I. D'Adamo, "The circular economy and bioeconomy in the fashion sector: Emergence of a "sustainability bias"," *Journal of Cleaner Production,* vol. 329, 2021.
39. E. M. Foundation, "Vision of a circular economy for fashion," 2020.
40. A. Fontana, A. Barni, D. Leone, M. Spirito, A. Tringale, M. Ferraris, J. Reis and G. Goncalves, "Circular Economy Strategies for Equipment Lifetime Extension: A Systematic Review.," *Sustainability,* 2021.

Critical Review of Circularity Indicators for Neighborhoods in Developing Countries

Nuri Cihan Kayaçetin

Abstract The construction industry is shifting from a conventional linear practice towards a circular economy sector. Led by the European Union (EU) Green Deal, initiatives such as the EU Circular Economy Action Plan, Renovation Wave, and 'Fit-for-55' package aim to minimize emissions and material use until 2050. Previous initiatives focused on the emissions related to the use phase of the built environment, so-called Scope 1 carbon emissions. On the other hand, several other emission sources exist in the whole life cycle of buildings and cities. Circularity can become a key element in mitigating the embodied carbon of building materials that originate from the production and end-of-life processes. Recently, cities and neighborhoods have been considered as suitable scale to develop and test innovation solutions. On the other hand, there is no consensus on measuring circularity at urban scale and most of the recent studies originate from the EU. Then, this study aims to explore urban circularity indicators in the context of developing countries. First, existing urban circularity indicators are reviewed and then two case study areas with different urban settings from Turkiye are selected for exploring the indicators. Several recommendations are provided for the utilization of indicators in distinct urban settings and climate regions.

Keywords Circularity · Environmental indicators · Neighborhood · Developing countries

1 Introduction

The construction industry is under pressure to provide innovative methods to achieve the goals introduced by the EU Green Deal [1]. While the Fit-for-55 package focuses on mitigating all carbon emissions by 55% [2], EU Circular Economy Action Plan has

N. C. Kayaçetin (✉)
Department of Interior Architecture and Environmental Design, Faculty of Art, Design and Architecture, Bilkent University, Ankara, Türkiye
e-mail: cihan.kayacetin@bilkent.edu.tr

A. Podleisek et al. (eds.), *Advances in Circular Economy: Bridging Research and Practice*, Springer Proceedings in Earth and Environmental Sciences,
https://doi.org/10.1007/978-3-032-21547-5_11

been introduced to accelerate the integration of circular methods in all industries [3]. The last but not the least, the Renovation Wave focuses on improving the performance of the existing building stock by tripling the deep renovation rate [4]. The overall goal is to achieve a climate-neutral industry by 2050. One of the challenges against this goal is mitigating embodied carbon emissions. Currently, most of the building materials continue their service life linearly and end up as down-cycled materials or waste. Several circular initiatives and principles have been developed to address the potential for improvement in production, replacement and end-of-life of products and buildings to change this behavior.

The most common framework that guides circular methods is based on the 9R principles of circularity that range from refuse, rethink, reuse to recycle [5]. The framework focuses on improving material efficiency by up-cycling materials at their end-of-life. Ultimately, it also responds to the depletion of raw materials and resources. There are many studies on the circularity of building materials and how such materials can be utilized by following 9R principles at a building level in the last decade. One of the pioneering studies is from Durmisevic [6] on design for disassembly (DfD) methods for buildings. Following the research on circularity principles, there has also been an emerging interest in bio-based building materials that champions over petrol-based materials due to their high capacity of dissolving or potential for re-use [7–9]. On the other hand, when the scale is increased to urban circularity, the amount of research is quite limited.

The definition of urban circularity is not clear. For some, it represents an urban area that does not generate waste. For others, it is an overarching framework that enables a city or a region to fully sustain itself. While the concept originates from an environmental perspective, it involves socio-economic aspects at an urban level. To provide a comprehensive definition, several researchers have worked on urban circularity indicators and methods as seen in Table 1.

The research in the last five years displays a strong research preference for urban circularity. Depending on the keywords provided by the authors, some common trends were identified: (i) assessment frameworks have a strong focus on cities, (ii) adoption of CE principles, and (iii) circularity indicators. Besides the circularity principles and frameworks, some authors explore bio-based building materials as a means to increase circularity, when compared to conventional construction materials. Mouton et al. [7] and Le et al. [8] investigated LCA and sustainability of bio-based building materials, with a focus on their reuse and recycling potential.

With regards to the development of assessment frameworks, Ghisellini et al. [10] and Vanhuyse [11] provided insights into circularity at the macro level, offering perspectives on circular urban metabolism and frameworks for planning, monitoring, and evaluating CE transitions in cities. The authors pointed out the need for comprehensive approaches to assess circularity that consider broader urban systems and dynamics. Additionally, there is an interest in integrating social aspects into circularity assessments. Vanhuyse et al. [14] and Cui [15] highlighted the inclusion of social factors in CE for understanding circularity initiatives on an urban scale. The authors agreed with the need for holistic approaches that include environmental and social assessment of urban circularity.

Table 1 Literature review on urban circularity

Ref	Author	Year	Keywords
[10]	Ghisellini et al.	2023	Macro-level CE, urban circularity, national indicators
[11]	Vanhuyse	2023	CE, circular cities, framework, indicators, measurement
[12]	Luciano et al.	2023	Urban mining, disused buildings, circularity indicators, reuse
[13]	Paoli et al.	2022	Urban sustainability, circular cities, indicators
[14]	Vanhuyse et al.	2022	CE, urban transitions, employment conditions, participation
[15]	Cui	2022	Circular urban metabolism, material flows, resource use patterns, urban systems, sustainability
[16]	Veeckman and Temmerman	2021	Citizen science, participation, ULL, flood risk management; urban cities, environmental monitoring, policy making
[17]	Ascione et al.	2021	CE, ULL, nature-based solutions, sustainable transition, Turin
[18]	Vanhuyse et al.	2021	CE, cities, evidence mapping, urban transitions, 10R
[19]	Superti et al.	2021	CE, construction and demolition, indicators, R-strategies
[20]	Stugholm	2020	Urban CE, urban metabolism, framework, indicator
[21]	Corona et al.	2019	Sustainability, circularity metrics, LCA, validity, products, MFA, recycling
[22]	Girard and Nocca	2019	CE, circular city, health city, impact indicators
[23]	Fratini et al.	2019	CE, urban sustainability transitions, socio-technical imaginaries, co-production, governance, geographies of transition Europe

When these studies were checked depending on the preferred indicators; Girard and Nocca [22] provided an extensive list of multi-dimensional indicators that were deducted from literature and official documents and categorized them under micro, meso and macro levels. Then, the authors proposed a specific list of indicators for port cities. Corona et al. [21] provided a critical assessment of circularity metrics depending on three main methodologies: LCA, MFA and input-output analysis. They evaluate the metrics on three pillars of sustainability. Paoli et al. [13] provided existing indicators for the circular city and proposed a set of 13 indicators categorized under economic, environmental, and social pillars. The authors also provided another categorization of existing indicators based on sector and country. For buildings, they identified 8 indicators with a focus on the ratio of buildings that follow R-principles. Vanhuyse et al. [18] reviewed studies on circular cities and identified key indicators used in the last decade. The authors were critical of the diversity and number of metrics employed in assessing urban circularity. Later, Vanhuyse [11] introduced the Urban Circularity Assessment Framework (UCAF), which provides a structured approach for planning, monitoring, and evaluating circular economy transitions in cities. The framework integrates a range of indicators (resource flows, circular economy practices, social dynamics, and policy frameworks) to provide a comprehensive assessment. The study was applied in two cities: Umea and Stockholm, Sweden.

On the other hand, it should be mentioned that the existing frameworks and the indicators they utilize have a strong dependence on the EU context. Hence, this study provides a critical review of existing metrics to identify proper indicators for urban circularity assessment in developing countries.

2 Method and Material

The methodology aims to adapt previous applications to the context of a developing country, Turkiye. As most studies were conducted in EU, there are environmental, social, economic and political differences. This study considers the environmental aspects of urban circularity. Another purpose of the study is to integrate building-level frameworks into urban-level frameworks. The assumption is that urban-level frameworks can benefit from building-level frameworks specifically on building material efficiency.

One of the prominent frameworks for building circularity belongs to Brand [24], who provided a shearing layer concept for the building materials depending on their service lifetime. The layers are ordered from the longest to shortest lifetime such as site, structure, skin, services and stuff. One of the first studies to adopt this framework is from Durmisevic [25] who focused on DfD building scenarios. On the other hand, when the system boundary is extended to an urban setting, some obstacles occur as elements beyond the building are either omitted or considered eternal (such as the site layer whereas the site or land does not have an infinite lifetime).

To understand and utilize resource flows and cycles between buildings and urban space, new indicators related to new urban layers should be introduced. Green and blue networks, supply grids, landscape, hardscape, and earth should also be categorized as a shearing layer of the built environment. The new layers allow the use of relevant indicators to understand the urban circularity concept. The exterior layer in building and urban space should be considered as the point of transition. Harmonizing technical and social indicators toward a comprehensive circularity assessment is an overall goal.

2.1 Indicators

To evaluate the circularity indicators, a criteria list from previous studies [20, 21] was adopted and categorized according to the pillars of sustainability as seen in Table 2. Then, the criteria related to environmental aspects were utilized to evaluate common circularity indicators. The economic benefits were neglected for the sake of simplicity.

Then, common circularity indicators were retrieved from several studies [13, 20, 22] and characterized according to the urban layers (Table 3). The criteria for the indicators are derived from Table 2 and adapted for the environmental aspects of circularity. The adaptation takes into account the life cycle phases of buildings; production and construction, use and maintenance phase and end-of-life. The indicators are marked if they have a positive impact on the respective criterion.

The indicators (I) 1-4 are related to the methods of construction and renewal of buildings. I1-2 represents the current trend in a linear construction industry whereas I3-4 represents a circular one. These indicators have a positive impact on extending building lifetime, but only through DfD methods, one can ensure the improvements are achieved through all life cycle phases of the building. I2 should be questioned whether it would have a direct impact on C4, increasing reused and recycled content. To ensure the link between deconstruction of buildings and rate of reused materials, I2 should be utilized together with I5.

I8-9 refers to the reduction of emissions levels during the use phase of the buildings. I1-4 have the potential to reduce emissions levels as well, but to ensure the increase in energy efficiency of buildings, special attention should be provided to I8-9. Currently, it is not possible to foresee the impact of circular processes on GHG emissions. As an example, reuse and recycling will definitely impact on the emissions of production facilities, but it does not necessarily have an impact on the neighbourhood where materials are reused. I6-7 refers to the utilization of green and blue networks for reducing resource depletion and increasing recycled resources.

Table 2 CE evaluation framework

	Criteria	Environment	Economic	Social
1	**Reducing resources depletion**	x		
2	**Reducing emission levels**	x		
3	**Reducing material waste**	x	x	
4	**Increasing renewable and recycled resources**	x		
5	**Extending product lifetime**	x	x	
6	Increasing local employment			x
7	Value-added creation and distribution		x	x
8	Increase social wellbeing			x

Adopted from Girard and Nocca [22]

Table 3 Categorization of urban indicators and evaluation criteria

Indicator	Layer	Criteria				
		C1 resource	C2 emission	C3 material	C4 recycled	C5 lifetime
1—Ratio of new buildings	Building		x			x
2—Ratio of degraded buildings	Building				x	x
3—Ratio of renovated buildings	Building	x	x	x		x
4—Ratio DfD buildings	Building	x	x	x	x	x
5—Reuse/ recycling of materials	Resource	x			x	
6—Recycling of wastewater	Blue	x			x	
7—Public green space per capita	Green	x				
8—Amount of energy circulated by on-site renewable energy	Energy		x			
9—Amount of GHG emissions avoided by circular processes	Energy		x			

In next sub-section, the indicators were discussed over two distinct case studies from Turkiye; an existing neighborhood in Menteşe district, Muğla and an urban park in Çankaya district, Ankara. The evaluation criteria guided the critical discussion in the Results section.

2.2 Case Study

The selected cases were characterized according to several parameters such as urban typology, user profile, challenges and climatic zones in Table 4. The differences between the two cases offer an opportunity to assess a variety of indicators.

Table 4 Case study properties

Parameters	Case I—Muğla Menteşe	Case II—Çankaya Bademlidere
Urban typology	Existing neighborhood	Urban park
Development type	Densification	Public space among urban sprawl
Area	285,000 m^2	750,000 m^2
Transformation	Re-construction	Social transformation
Construction method	Reinforced concrete (RC)	RC + landscape design
User profile	Low-mid income	Low and high-income
Urban challenges	Seasonal floods and droughts	Drought, social inequality
Climatic zones	Mediterranean	Continental

Aerial maps of the cases are provided in Fig. 1. Case I, Muğla Menteşe is an existing neighborhood in an area of 285,000 m^2 in the southwest region of Turkiye. There is a high rate of migration to the city due to socio-economic conditions and earthquake risks in other regions. The building stock is older than 30 years where renovation needs are as crucial as the demand for new construction. Most of the buildings are residential whereas the public buildings are rather new. Even though water storage is mandatory on a lot larger than 2000 m^2, there is an increasing demand for the infrastructure.

Case II is a newly developed urban park on 750,000 m^2 area in the capital city Ankara which may serve as a testing zone for infrastructural solutions. The urban open spaces offer a suitable area for off-grid resource generation and storage (energy, water, etc.). The neighborhood is hosting users from significantly different socio-economic backgrounds.

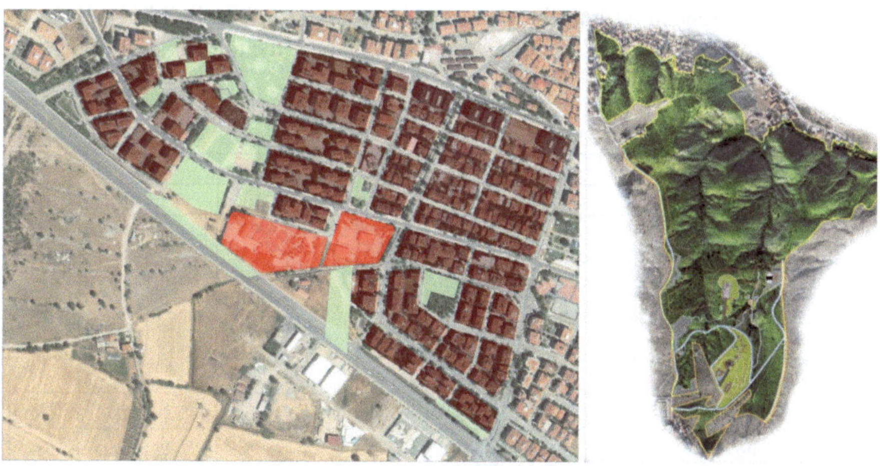

Fig. 1 Case I Muğla Menteşe (on left) and Case II Çankaya Bademlidere (on right)

In both cases, there are several common challenges such as immigrants from different regions of Turkiye, low purchase power due to the economic crisis, and urban sprawl due to high demand for accommodation and commercial areas. It should be highlighted that the awareness of environmental aspects is relatively lower than in EU.

3 Results and Discussion

In this section, each common urban circularity layer is discussed by giving examples of how they can be applied to different cases and what kind of results should be expected in the context of a developing country. First, the indicators that will significantly impact each case study (CS) depending on parameters are provided in Table 5.

Indicators related to buildings categorize the construction activities into new construction, degraded buildings, DfD buildings and the number of renovations. For Case I where housing demand is at peak level, the current construction figures will not be useful, as degrading and new construction are the main trend of the urban transformation. It is seen that:

- Structural renovation of existing buildings is not always possible due to high earthquake risk and low construction quality. Renovation is economically not feasible.
- Infrastructure does not allow for extending the urban zone hence densification is a possible choice. Indicators should be coupled such as 'reuse/recycling of materials' and 'public green space per capita' to evaluate the quality of such densification.
- 'Number of renovated buildings' may be limited to public buildings since the construction quality does not allow renovations to be technically possible.

Table 5 Parameters—indicators matrix

Indicators	Buildings					Blue and green		Energy	
Parameters	I1	I2	I3	I4	I5	I6	I7	I8	I9
Urban typology Development type Area Transformation Construction method	CS1	CS1	CS1	CS2	CS2	CS1	CS2	CS1	CS2
User profile	CS2	CS1	CS1	CS2	CS1	CS1	CS2	CS1	CS2
Urban challenges	CS1	CS1	CS1	CS2	CS2	CS2	CS1	CS1	CS2
Climatic zones	CS2	CS1	CS2	CS2	CS1	CS2	CS1	CS1	CS2

Indicators related to resources including blue and green networks hold great importance in the planning of new urban developments. In both cases, water management is quite critical; Case I experiences seasonal droughts in summer and floods in winter whereas Case II aims to perform as a resource hub for nearby neighborhoods. There is a lack of green and blue networks in urban areas. Increasing the ratio of blue and green networks is critical for Case I: to avoid over-densification and to provide local space for water storage, and for Case II: to avoid privatization of the public space and increase the water capacity.

Indicators related to energy and GHG emissions should be considered separately. Developing countries generally have carbon-intensive energy production. Hence, a neighborhood in Turkiye may save more GHG emissions than those in EU with the same amount of energy mitigation. Hence, it is suggested to focus on GHG savings to maximize the impact of circular processes. For Case II, it is important to note that urban parks can become an energy hub for nearby neighborhoods. The area can support cheap and low-carbon energy to users from various socio-economic backgrounds. One obstacle in developing countries is the lack of required infrastructure and legal basis to share the on-site generated energy in the immediate vicinity. Currently, such initiatives must sell the generated energy to the grid.

Depending on the discussion, the indicators are prioritized in Table 6 according to each case study to provide a general overview of the relevant layer of built environment and criteria. Prioritization was conducted via expert opinion gathered from local authority in the regions. Some of the indicators are disregarded due to the factors related to the current needs of users and trends of the construction industry. For Case I, the DfD buildings are not realistic due to lack of infrastructure and obstacles in local regulations (specifically for fire resistance and use of secondary materials in new construction). For Case II, the aim is to preserve urban public areas hence new construction or de-construction is not a priority. They would benefit from the indicators that are related to energy and resource efficiency.

4 Conclusion

The local context of a neighborhood should be considered as the baseline to initiate the circular economy implementations. As seen in the provided cases, some common indicators may not provide expected outcomes in developing countries due to socio-economic or technical obstacles. Hence, the 9R principles of circularity were considered in a stepwise approach for neighborhoods in developing countries to improve gradually.

As the number of circular cities increases, the literature will provide more guidance on how to tackle the local problems within a general framework. As a conclusion of this study, it is advised to cluster circularity indicators into different phases of 9R principles depending on the local context to (i) avoid obstacles such as lack of data, over-ambitious targets and (ii) achieve realistic targets with the local authorities. It is relatively easy to 'refuse and rethink' in an urban public space (Case II) but it may

Table 6 Prioritization of urban indicators for selected case study locations

Indicator	Layer	Criteria	Case I	Case II
			Menteşe	Bademlidere
Ratio of new buildings	Building	2, 5	1	–
Ratio of degraded buildings	Building	4, 5	2	–
Ratio of renovated buildings	Building	1, 3, 4, 5	6	7
Ratio DfD buildings	Building	1, 2, 3, 4, 5	–	5
Reuse/recycling of materials	Resource	1, 4	5	4
Recycling of wastewater	Blue	1, 4	7	3
Public green space per capita	Green	1	4	2
Amount of energy circulated by on-site renewable energy	Energy	2	3	1
Amount of GHG emissions avoided by circular processes	Energy	2	–	6

be better to 'repurpose' in an existing area (Case I). For existing neighborhoods in a developing region, the ratio of reused materials and green space per capita may be more important than the number of DfD or renovated buildings. For municipalities, a framework with several phases depending on the maturity of circularity could be a feasible solution.

The future studies should include a critical review of economic and social indicators. This study focused on technical aspects of circularity. Then the indicators may be validated for replication in other regions. This would require a multi-disciplinary approach which suits the concept of circular economy. There is a need to increase the number of case studies in developing countries to further test recommended metrics in literature.

Acknowledgement This research is a deliverable of the project Urban-CoLLaR: Co-creating Urban Living Labs for Circular Regenerative Neighbourhoods (F-DUT-2023-0138). This project was funded by the Scientific and Technological Research Council of Türkiye (TÜBİTAK) (Project nr:224N048), Ministry of Enterprises and Made in Italy (MIMIT) (P/130001/01-02/X73), The Research Foundation—Flanders (FWO) (Project nr: S010323N), Taskforce for Applied Research (DUT.02.2), The National Centre for Research and Development (NCBR) (DUT/2023/4/Urban-CoLLaR/2025), and Executive Agency for Higher Education, Research, Development and Innovation Funding (UEFISCDI) (Contract no 121 from 20/02/2025), as part of Driving Urban Transitions (DUT) 2022-2025, co-funded by the European Union's Horizon 2020 research and innovation programme under Grant Agreement No. 101069506.

References

1. EC. Communication (EU) 2019/640 from the Commission to the European Parliament, The Council, The European Economic and Social Committee and the Committee of the Regions—The European Green Deal. 2019. 2019.
2. Wilson A, Danti Renew N, Jutta Paulus I. BRIEFING EU Legislation in Progress Proposal for a Directive of the European Parliament and of the Council on energy efficiency (recast) Revising the Energy Efficiency Directive: Fit for 55 package. 2022.
3. EC. Communication (EU) 2020/98 from the Commission to the European Parliament, The Council, The European Economic and Social Committee and the Committee of the Regions—A new Circular Economy Action Plan for a cleaner and more competitive Europe. 2020.
4. EC. Communication (EU) 2020/662 from the Commission to the European Parliament, The Council, The European Economic and Social Committee and the Committee of the Regions. A Renovation Wave for Europe—greening our buildings, creating jobs, improving lives. 2020.
5. Ellen MacArthur Foundation. Growth Within: A Circular Economy Vision for a Competitive Europe. 2015.
6. Durmisevic E, Yeang K. Designing for disassembly (DfD). Architectural Design. 2009.
7. Mouton L, Allacker K, Röck M. Bio-based building material solutions for environmental benefits over conventional construction products—Life cycle assessment of regenerative design strategies (1/2). Energy Build. 2023;282:112767.
8. Le DL, Salomone R, Nguyen QT. Circular bio-based building materials: A literature review of case studies and sustainability assessment methods. Build Environ. 2023;244:110774.
9. Kayaçetin NC, Verdoodt S, Lefevre L, Versele A. Integrated decision support for embodied impact assessment of circular and bio-based building components. Journal of Building Engineering. 2023;63.
10. Ghisellini P, van Langen SK, Patil RA, Ramakrishna S. Circularity at Macro Level: The Urban and National Perspectives. Circularity Assessment: Macro to Nano. Springer Nature Singapore; 2023. p. 37–55.
11. Vanhuyse F. The Urban Circularity Assessment Framework (UCAF): a Framework for Planning, Monitoring, Evaluation, and Learning from CE Transitions in Cities. Circular Economy and Sustainability. 2023;
12. Luciano A, Altamura P, Baiani S, Cutaia L. The building stock as an urban mine: The case of the circular regeneration of disused buildings. Sustain Chem Pharm. 2023;33.
13. Paoli F, Pirlone F, Spadaro I. Indicators for the Circular City: A Review and a Proposal. Sustainability (Switzerland). 2022;14.
14. Vanhuyse F, Rezaie S, Englund M, Jokiaho J, Henrysson M, André K. Including the social in the circular: A mapping of the consequences of a circular economy transition in the city of Umeå, Sweden. J Clean Prod. 2022;380.
15. Cui X. A circular urban metabolism (CUM) framework to explore resource use patterns and circularity potential in an urban system. J Clean Prod. 2022;359.
16. Veeckman C, Temmerman L. Urban living labs and citizen science: from innovation and science towards policy impacts. Sustainability (Switzerland). 2021;13:1–15.
17. Ascione GS, Cuomo F, Mariotti N, Corazza L. Urban Living Labs, Circular Economy and Nature-Based Solutions: Ideation and Testing of a New Soil in the City of Turin Using a Multi-stakeholder Perspective. Circular Economy and Sustainability. 2021;1:545–62.
18. Vanhuyse F, Haddaway NR, Henrysson M. Circular cities: an evidence map of research between 2010 and 2020. Discover Sustainability. 2021;2.
19. Superti V, Houmani C, Binder CR. A systemic framework to categorize Circular Economy interventions: An application to the construction and demolition sector. Resour Conserv Recycl. 2021;173:105711.
20. Stugholm S. Developing an Urban Circular Economy Framework Based on Urban Metabolism. 2020.

21. Corona B, Shen L, Reike D, Rosales Carreón J, Worrell E. Towards sustainable development through the circular economy—A review and critical assessment on current circularity metrics. Resour Conserv Recycl. 2019;151:104498.
22. Girard LF, Nocca F. Moving towards the circular economy/city model: Which tools for operationalizing this model? Sustainability (Switzerland). 2019;11.
23. Fratini CF, Georg S, Jørgensen MS. Exploring circular economy imaginaries in European cities: A research agenda for the governance of urban sustainability transitions. J Clean Prod. 2019;228:974–89.
24. Brand S. Shearing Layers of Change. How buildings learn: What happens after they're built. 1994.
25. Durmišević E. Reversible building design. Designing for the Circular Economy. 2018.

Advancing Circular Economy in Fashion: Feasibility and Impact of Enhanced Recycling Technologies for Post-industrial Textile Waste

Marco Ricchetti, Aurora Magni, Alessandro Fontana, Deborah Leone, and Ludovica Rossi

Abstract The circular economy is crucial for the contemporary fashion industry, as it offers a way to tackle the numerous environmental and sustainability issues associated with the traditional linear approach to fashion production and consumption. A paramount challenge lies in drastically advancing recycling practices within the entire fashion value chain and managing the garments' end-of-life. Such advancements would allow the industry to recover the value of materials currently wasted throughout the journey from raw materials to garments in stores. The paper summarizes the exploration of the feasibility and the potential impact of the adoption of recycling technologies in various real-case scenarios, and it is based on primary data and information collected in (a) a field research conducted by Blumine in the framework of a UNIDO SwitchMed project focusing on the valorisation of post-industrial textile waste in a circular economy perspective in three North-Africa countries (Egypt, Tunisia, and Morocco), (b) in the evaluation of the environmental impact of new technologies for the treatment of textile fibres blended with elastane. Both (a) and (b) studies have highlighted the need to address the issue of the recycling of mixed fibres that represent a large share of the textile waste

Keywords Circular economy · Textile recycling · Sustainable fashion

M. Ricchetti (✉) · A. Magni
Blumine Srl, Milan, Italy
e-mail: marco.ricchetti@supsi.ch

A. Fontana · D. Leone · L. Rossi
University of Applied Science of Southern Switzerland, Viganello, Switzerland

A. Podleisek et al. (eds.), *Advances in Circular Economy: Bridging Research and Practice*, Springer Proceedings in Earth and Environmental Sciences,
https://doi.org/10.1007/978-3-032-21547-5_12

1 Introduction

Northern African Mediterranean countries have emerged as a significant production hub for garments. Four countries stand out as major suppliers in the region: Turkey, Egypt, Tunisia, and Morocco, collectively exporting around two million tons of textile garments, predominantly to the EU [1].

The substantial textile and clothing export and production activities in these countries result in significant amounts of textile waste. Research carried out by UNIDO from 2020 to 2023 in Egypt, Tunisia, and Morocco [2–4], examined the post-industrial waste produced at various stages of the supply chain, from spinning and weaving through to the creation of garments, including the disposal of unsold or "second quality" items (e.g., the pre-consumer textile waste flows generated by Morocco's textile and clothing industry are estimate to be 83,200 tons per year [5]).

In assessing the factors that facilitate the creation of a waste valorisation supply chain aligned with circular economy (CE) principles, the presence of relevant technologies and the country's regulatory and legal structures governing waste management play crucial roles [6]. The recycling infrastructure in the three countries is notably inadequate and fails to keep pace with recent trends in innovation. Additionally, the existing waste management regulations require modernization to effectively incorporate CE principles [2–4]. The opportunity to reduce environmental impacts, particularly in terms of greenhouse gas emissions and resource depletion, through national CE strategies in these countries is significant [7, 8].

Given that a large portion of these countries' garment production is destined for the European market, implementing such strategies could also play a crucial role in lessening the environmental footprint of European consumption.

The paper addresses the following research questions:

- Are recycling technologies economically sustainable?
- Are recycling technologies environmentally sustainable?
- Should the technologies be complemented by other enabling elements, such as regulations and training for companies and their suppliers, to encourage waste impact reduction?

2 Background: Overview of State-of-the-Art Technology for Mixed Fibres Recycling

Textile recycling technologies are taking centre stage as the demand for circular polymers rapidly increases [9]. New advanced textile recycling technologies have been developed, especially in Europe, driven by the adoption of specific regulations [5] and the issuance of strategic documents by the European Commission focusing on CE [7]. The recycling routes for textile waste recycling processes can be grouped into three main technology types [8]: (i) mechanical, (ii) thermo-mechanical, (iii) chemical recycling.

Mechanical Recycling (MR) involves waste shredding and tearing processes that result in spinnable fibres, fluff material or flocks for non-woven production, fillers, and additional waste streams of non-textile materials. MR has been the predominant approach for several decades and remains effective with high-quality, relatively clean, and sorted waste. However, it faces structural limitations such as limited pools of appropriate feedstock, degradation of recycled fibres, and other material properties that restrict end-market applications [10].

Thermo-Mechanical Recycling (TMR) is applicable to synthetic fibres and employs heat to melt thermoplastic polymers. Waste consisting of 100% pure synthetic fibres is ground and melted to produce chips, which are then used for extrusion to create new synthetic staple fibres or filaments. However, this technology is restricted to feedstock comprising 100% single thermoplastic fibres and may be constrained by the presence of unwanted chemical auxiliaries and contaminations [11].

Chemical Recycling (CR) comprises new and advanced technologies applicable to synthetic or cellulose-based polymers (including cotton, which can be converted into cellulose) and material blends. Fibers undergo recycling through chemical reactions that break down the polymeric chains, re-creating the original monomers for reuse in creating new polymeric materials. Cellulose-based polymers are typically recycled by extracting cellulose pulp from the fibre and dissolving it in a solution that enters the paper supply chain or can be transformed into viscose fibre for textile processes. However, current capacity remains limited, with many CR technologies still in the developmental and scaling phases [12].

CR offers several advantages over MR: (i) it remains the sole option for effectively recycling mixed fibres, as well as degraded or contaminated polymers and heavily damaged fibres; (ii) it enables the production of recycled fibres with technical and quality characteristics akin to virgin fibres, making them suitable for multiple recycling cycles. This virgin-like quality is a significant advantage of CR over MR, which typically yields lower quality and contaminated recycled fibres. Consequently, recycled fibres often need to be blended with virgin materials to achieve yarn of acceptable quality and cannot typically be recycled at end-of-life without mixing them with virgin materials; (iii) it allows for greater flexibility in achieving desired quality, colour, and performance in the final product. In contrast, MR is unable to eliminate (hazardous) chemicals such as additives, dyes, finishes, etc., that may be present in textile products throughout production and product use, resulting in these chemicals remaining in the output [12].

Conversely, while CR presents several benefits, it is not without its drawbacks. These include the requirement for higher initial investments to set up the necessary infrastructure and technology and is characterized the higher energy intensity, which, in turn, can lead to increased GHG emissions posing significant challenges to the scalability and environmental sustainability of CR methods. Furthermore, the Technology Readiness Level (TRL) of MR and TMR technologies are more advanced, compared to CR ones that are still in the developmental phase and have not yet achieved commercial readiness or widespread market adoption at a large scale [8]. However, technological innovation is rapidly reshaping the recycling landscape. After a century-long reliance on MR for natural fibres and several decades of progress in TMR for thermoplastic (synthetic) fibres, CR technologies are currently at the forefront of innovation, with significant advancements anticipated in the near future [11].

In this context, the CR potential for the improvement of the environmental performance and scalability of recycling technologies makes it a key enabler of circularity in the textile industry.

An example of the benefits/trade-off of CR is found in textile fibres dissolution methods. These processes rely on solvents, which can be toxic, flammable and contribute to air pollution. Adopting safer and biobased solvents offers a more sustainable alternative that reduces drawbacks. On the other hand, this solvent-based procedure can save 60.3% CO_2-eq./kg of textile waste compared to incineration. Thus, the developed recycling process can have a positive impact on the environment by reducing overall greenhouse gas emissions and can increase the resource circularity in the textile industry.

2.1 Overview of the New Chemical Recycling Technologies

CR of textile fibres is not a monolithic field. It encompasses a range of new and advanced technologies applicable to synthetic or cellulose-based polymers (including cotton, which can be converted into cellulose) and material blends [13]. Fibers are recycled through chemical processes that either break down the polymer chains into their original monomers, which can then be reused to manufacture new polymer materials, or purify-dissolve the unwanted chemical components, leaving behind pure fibres that can be reprocessed.

Specifically, the study addresses two main approaches to CR: (i) monomer recycling (or depolymerization), and (ii) polymer recycling (or solvent-based purification).

Monomer Recycling (*Depolymerization*) involves breaking down polymeric fibres into their constituent monomers and re-polymerizing them for new uses. Depolymerization processes may employ different types of chemical reactions, such as hydrolysis using water, methanolysis using alcohols, or glycolysis using glycol.

Polymer Recycling (*Solvent-Based Purification*) focuses on purifying and extracting polymers from mixed textile waste for reuse. The purification process utilizes various techniques depending on the type of feedstock being treated. For instance, polycotton blends are primarily recovered through solvent-based dissolution and filtration processes (solvent-based purification), hydrothermal approaches, or enzymatic routes (biochemical recycling). The concept of purification or dissolution can be applied to mixed materials involving synthetic or cellulosic fibres or to pure cellulosic fibres. Typically, cellulose-based waste, such as cotton waste can be recycled by extracting cellulose pulp from the fibre and dissolving it in a solution, which can either enter the paper supply chain or be transformed into viscose fibre for textile processes.

Currently, the production capacity for CR of textile waste remains limited, and most CR technologies, in both the depolymerization and the solvent-based purification, are still in the developmental and scaling phases [8]. CR technologies for plastic packaging, such as PET or Polyamide, which necessitate a cleaner and more selectively chosen feedstock, are, however, at a more advanced stage of development.

3 Methodology: Economic Feasibility and Impacts Evaluation

3.1 A Feasibility Study for a Chemical Recycling Plant in Egypt

A feasibility study for a large-scale investment in a CR plant in Egypt was conducted by Blumine and CSIL Milano as part of the UNIDO SwitchMed project [2–4] aimed at promoting the development of a circular value chain for the textile industry in Egypt.

Egypt's textile and clothing industry generates significant post-industrial and pre-consumer textile waste from producing textiles and garments consisting of cutting scraps, second-quality products, spillage, and other textile waste. Almost half the waste is mixed fibres that can hardly be recycled back to fashion.

This waste amounts to approximately 212 kton, according to [2], out of which more than 50% is composed of either 100% cotton or cotton-rich blends. However, nearly half of the textile waste generated, approximately 44%, consists of materials that, although typically downcycled or incinerated, with consequent economic and environmental costs, can potentially be recycled via CR technologies. These materials include 100% polyester, poly-rich blends, other synthetic fibres, and mixed synthetics that may encompass acrylic or nylon, cotton-elastane, viscose, lyocell, acetate, and more.

Given the importance of synthetic fibres and blends in the composition of textile waste in Egypt, national stakeholders aim to define utilization opportunities for textile waste made of polyester, poly-rich blends, and other synthetic blends.

In this respect, MR could be, in principle, considered as a potential option, not only because it has already reached TRL 9 but also because Egypt already features some MR capacity both for cellulosic and synthetic fibres [2].

However, CR holds significant promise in Egypt due to its advantages over other technologies, including greater flexibility in the range of feedstock fibres, the virgin-like quality of chemically recycled fibres, and enhanced flexibility in achieving desired quality, colour, and performance in the final product.

3.2 The Methodology of the Feasibility Study

The study was undertaken by a team comprising experts from Blumine and CSIL Milano, along with cooperation from technology providers of CR solutions: GR3N and Worn Again, as well as the Egypt Ministry of Industry through its specialized agencies, the Industrial Development Agency and the Industrial Modernization Center.

The investigation focused on the feasibility of two CR approaches: hydrolysis-based depolymerization and solvent-based purification, chosen for their specialization by two European technology providers. The UNIDO Manual for preparing feasibility studies [13], a standard for conducting feasibility studies, served as the principal guideline for data gathering and the drafting of this feasibility study.

The feasibility study scenario accounted for the recycling process producing two types of recycled materials: rPET (recycled Polyethylene Terephthalate, often called Polyester in textile terminology) and cotton pulp, which can be utilized to manufacture viscose (see Fig. 1).

Central to the study was a set of assumptions and parameters, forming the basis of a financial analysis of their relative economic viability and constraints of both recycling options. Each recycling method displays distinct advantages, notably in the variety of materials they accept, their efficiency, the minimum operational scale of a plant, and the nature of the end products.

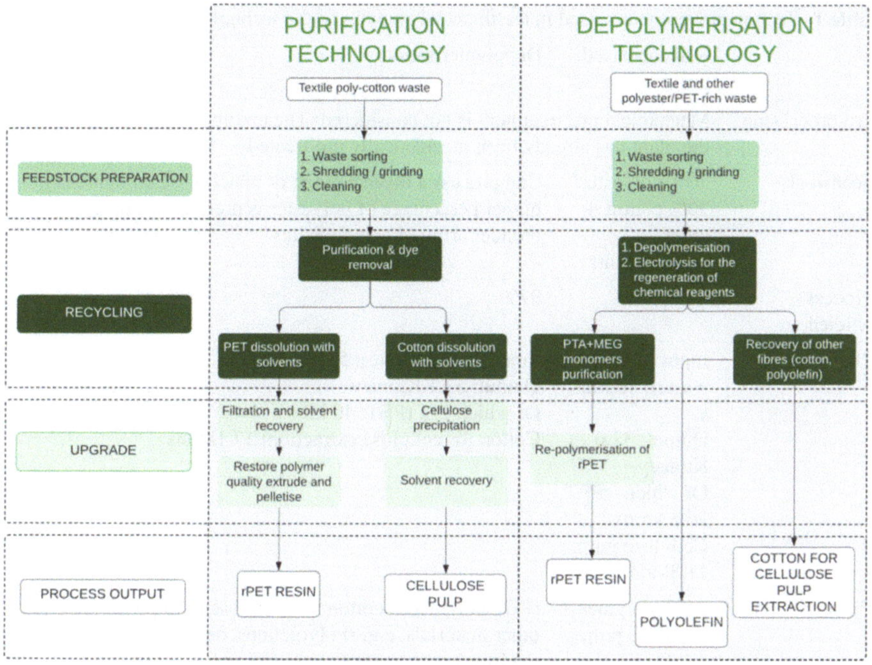

Fig. 1 Flow chart of solvent-based purification and depolymerization recycling processes. *Source* [14]

The analysis considered specific assumptions regarding the composition of recycled materials and the stages of the process evaluated for financial and operational impact. The same material blend—70% polyester and 30% cotton—was assumed for both technologies, although they are capable of processing various textile blends.

The solvent-based purification technology shows a narrower flexibility regarding material composition, suitable for processing blends like 65% polyester and 35% cotton, with a primary focus on these fibers. Its tolerance for other fiber types, such as polyamide or polypropylene, and contaminants is limited, accepting only minimal quantities that are not reclaimed in the process.

Conversely, the depolymerization approach offers greater versatility, accommodating a wide spectrum of synthetic materials, including polypropylene and polyethylene, which are recuperated as polyolefin. Moreover, it can handle different proportions of cellulosic fibers and a wider range of contaminants (see Table 1).

The financial analysis utilizes the discounted cash flow method that estimates the value of an investment using its expected future cash flows [15]. The operational guidelines established are as follows:

- Only cash inflows and outflows were accounted for, excluding non-cash items such as depreciation and amortization, reserves, price and technical contingencies, and other accounting entries not representing actual cash movements.

Table 1 Technical parameters used in the financial analysis of the technologies [14]

	Solvent-based purification	Depolymerisation
Pre-processing	Mechanical pre-treatment is not considered: The textile waste used as input in the plant has already been mechanically pre-treated	
Feedstock	70% polyester/ 30% cotton + other contaminants	Can process a diverse range of materials. Economically, a higher percentage of polyester is preferable. Accepts a broader array of contaminants
Process efficiency	94%	97%
Plant capacity	Input (textile waste): 40 tons/ y Output: 37.6 Ktons/y Of which: rPET: 26.3 ktons Cellulose pulp: 11.3ktons	Input (textile waste): 59 Ktons/y Output: 57.1 Ktons/y Of which (*): rPET: 40 ktons Cotton for cellulose extraction: 17.1ktons
Final product	rPET polymer + cellulose pulp	rPET polymer + cotton for cellulose pulp production + other materials, e.g. rPolyolefines, depending on the feedstock composition

- Taxes were excluded from the analysis to assess the project's commercial viability independently of its financing or incentives (i.e., without considering who pays what and how).
- The analysis was conducted in euros, using constant 2023 prices. Inflation was not factored into the analysis due to the difficulty in pinpointing specific cost items that might be affected by varying inflation rates.
- A reference period of 20 years was chosen to align the project's cash flow forecasts with its economically useful life and projected long-term impact. This period, spanning from 2023 (the base year) to 2043, includes six years for construction and pilot production, followed by 14 years of full production capacity, as detailed in the implementation plan. This plan was reconstructed based on data from European technology providers and local stakeholders' input on the permit acquisition timeline.
- The Weighted Average Cost of Capital (WACC) was applied as the discount rate to calculate the project's return on investment from potential investors' perspective. The real WACC used was 15%, derived from the nominal WACC (27%, as reported by Commercial International Bank Egypt in June 2023) minus the expected inflation rate (12%, estimated from Central Bank of Egypt historical inflation data and International Monetary Fund inflation forecasts).

The study evaluated the investment options' profitability by examining the capacity of future net revenues to cover the initial investment, excluding financing sources or methods and before considering incentives, along with the operational costs of running the plant at the desired scale. These financial elements—revenues, initial investment, and operating costs (excluding incentives)—were used to determine net cash flows before taxes and without amortization.

Three key indicators were calculated to assess the profitability of the two investment options:

- Internal Rate of Return (IRR): the rate that equalizes the net present value of cash inflows (revenues) and outflows (capital and operating expenses) to zero. The project is deemed commercially viable if the IRR exceeds the 15% financial discount rate.
- Gross profit margin: the profit percentage relative to generated revenue or sales. A positive margin indicates potential commercial viability.
- Payback period: the duration required for the project to recover its initial investment. A shorter payback period indicates a quicker investment return. This can be evaluated over the entire 20-year horizon or from the start of pilot production (year 6). Note that the payback period is distinct from the 20-year investment horizon. The former refers to the time needed to fully recoup the investment, significantly shorter than the investment timeframe. The latter denotes the "standard" operational lifespan, years until plant replacement, which influences the calculation of amortization costs (the shorter the lifespan, the higher the amortization cost).

CR technologies hold significant promise especially for Egypt, thanks to the country's extensive pool of feedstock compared to Tunisia and Morocco. The feasibility study found that investment in a CR plant in Egypt—estimated to cost around EUR 100 million for solved-based purification technology and between EUR 150 and 200 million for depolymerization technology—will see its profitability heavily influenced by the future trajectory of recycled PET (rPET) prices, which have been notably volatile in recent years. According to the scenario outlined in the study, the investment yields a positive gross profit margin when the price of recycled PET (rPET) exceeds €900 per ton for depolymerization and is approximately €1200 per ton for solvent-based purification. The IRR surpasses the weighted WACC value of 15% when the rPET price reaches €2000 per ton for depolymerization and €2100 per ton for solvent-based purification. Within the context of Egypt's economy, this gives depolymerization technology a competitive advantage over solvent-based purification technology.

3.3 The Methodology of the Environmental Impact Assessment

Within this framework, the environmental impacts of the new recycling technology have been evaluated using the Life Cycle Assessment (LCA) method. A comparison of the environmental footprint between the standard production process with a new process was studied. In this specific case, a virgin cotton product that is sent to incineration at the end of its product lifespan is compared with a product that uses partly recycled materials. The purpose of this study is to determine whether the new recycling process offers environmental advantages over the standard process currently used. The reference system, i.e. a t-shirt, was modeled according to the specific case of a major textile company. A cradle-to-grave approach is the selected approach for the LCA as it allows for realistic analysis of the whole life cycle of the product, from material extraction to end of life phase. The functional unit adopted is 0.2 kg of colored t-shirt.

The comparison involves the use of carded cotton to produce garments without the elastomer addition. In this case, the first scenario (scenario 1) considers the use of 100% virgin carded cotton whereas the second scenario (scenario 2) considers the use of 30% by weight virgin carded cotton and 70% recycled cotton. The recycled cotton is obtained through the innovative CR process, followed by the MR process (made possible by the elimination of the elastomer from the fibers recovered from the end-of-life). In both scenarios, an end-of-life (EoL) via incineration is assumed. This case analyzes the benefits of the CR and MR processes compared to the use of virgin material.

An overview of this case study is graphically provided in Fig. 2.

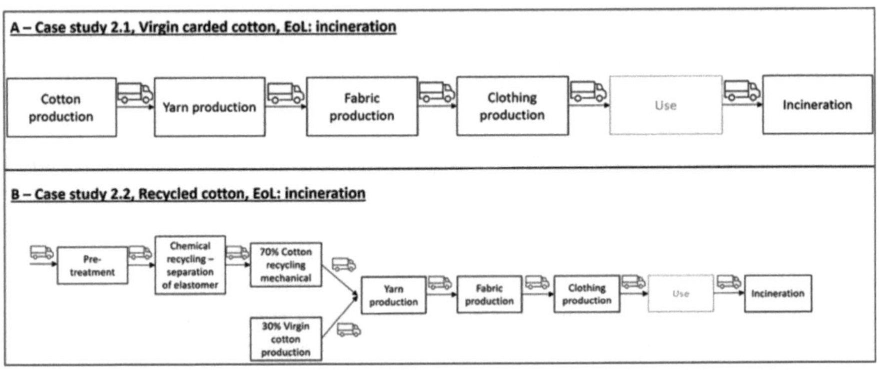

Fig. 2 Overview case study 2. **a** Process flowchart describing scenario 1 which has virgin carded cotton in input. **b** Process flowchart describing scenario 2 which has recycled and virgin cotton in input

4 Results: Investment Opportunities in Advanced Textile Waste Recycling Technologies across Three North African Countries

4.1 The Profitability Analysis

A profitability analysis was carried out by Blumine in Egypt, Tunisia and Morocco within the framework of the UNIDO SwitchMed project for investments in cutting-edge MR technologies. In Morocco and Egypt, the scenario was developed for 100% use of the recycled Textile-to-Textile (T2T) applications where the recycled fibre is used to make new yarns in blend with virgin cotton (30% recycled 70% virgin). In Tunisia a scenario was also tested with a blend of end uses in T2T and nonwovens applications.

The analysis evaluated as profitable the investment in modern MR equipment in all three countries factoring in the local operational (labour, energy, and feedstock costs) and land and building costs.

The calculation of production costs revealed that the largest component of production cost is the amortization of capital, accounting for approximately half of the cost per kilogram of recycled fibre produced.

Scenario A, which involves the production of higher-quality recycled cotton fibre for spinning, showed similar outcomes for Egypt and Tunisia (see Table 2). Available information indicated higher prices for both the feedstock and the recycled fibres in Morocco. Generally, the investment cost is expected to be recouped within 4–5 years. The situation differs for Scenario B, focusing on the production of lower-quality recycled cotton fibre for nonwoven applications, where Tunisia exhibited lower profitability and a longer payback period.

The study revealed a shared structural deficiency across the three countries, resulting in extremely low and ineffective post-industrial waste collection rates. This challenge perpetuates a vicious cycle, driven by the minimal demand within the local recycling value chain. This situation also presents a paradox where these countries import recycled fibres or yarns containing recycled materials yet fail to collect domestic post-industrial waste.

4.2 External Factors: The Legal Framework for Waste Management in the Three Countries

Finally, the analysis of the legal framework regarding waste management has found disparities among the three countries. In Egypt, a new law focusing on waste management was passed by the parliament in 2020, and the relevant executive regulation was issued in 2022. The New Law[1] provides an integrated framework covering a wide

[1] Law No. 202/2020 and Prime Minister's Decision No. 722/2022.

Table 2 Investment in a modern MR unit

Country	Yearly production volume (tons)	Target recycled fibre selling price (kg)	EBITDA/ sales (%)	EBIT/sales (%)	Payback period (years)
Egypt, scenario A	1.200	€0.8	60.9	26.4	4
Tunisia, scenario A	1.200	€0.8	59.6	17.2	5
Tunisia, scenario B	1.600	€0.4	50.0	0.9	7
Morocco, scenario A[a]	1.200	€1.2	40.2	14.1	4

Profitability indicators
Scenario A. production of higher quality recycled cotton fibre for spinning (T2T)
Scenario B. production of lower quality mixed fibre for nonwoven applications
[a] In the original study for Morocco, the construction cost for the building was not included, as the company already possessed the required space. However, for the sake of uniformity in this table, the amortization cost for the building has been added. *Source* Blumine [5]

range of subjects, including the collection, storage, sorting, transportation, treatment, valuation, recycling, reuse, and disposal of waste, all through environmentally safe standards. The new law aligns with the European legal standards, incorporating key principles such as "the polluter pays" and "Extended Producer Responsibility." It introduces a "Green Label System" aimed at promoting product recyclability and sets up a national waste register as well as a system for accrediting waste management operators. Nonetheless, this law is in the early stages of implementation, and its true effects will become apparent in the forthcoming years.

In Tunisia a structured system of licensing and regulations regarding waste management exists,[2] that nonetheless does not reflect the principles of CE and needs modernization. Moreover, the knowledge of the law is very limited among the business community, leading to a lack of awareness about the opportunities.

In Morocco, notwithstanding some later modifications, the structure of the legal framework for waste management is still based on Law No. 28–00, introduced 24 years ago and primarily concerned with safe dumping of hazardous waste in landfills. A structured licensing system or a national of waste are also missing.

4.3 Sustainability Impact Assessment Results

Considering the circularity of the recycling process, a significant reduction in environmental impacts using recycled cotton was demonstrated in Fig. 3. The use of recycled cotton reduces the environmental impacts in Global Warming Potential

[2] Law 1996–41, amended by Law 2001–14.

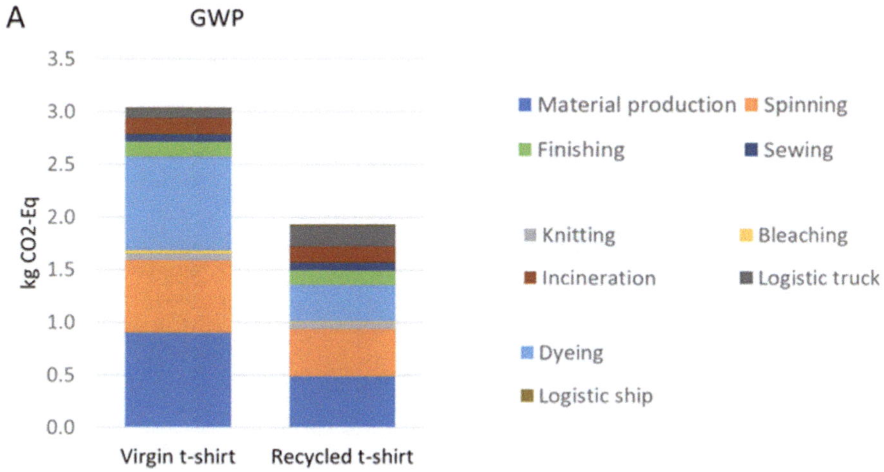

Fig. 3 Stacked bar chart showing GWP impacts results for the case study

(GWP) indicator by 36% compared to the linear case. Detailed results showed an impact of 3.04 kg CO_2-eq for scenario 1, and 1.93 kg CO_2-eq for scenario 2.

Overall, recycling strongly decreases the environmental impacts of cotton cultivation and dyeing, which are the process steps with the highest contribution to environmental impacts (Fig. 3). Additionally, the approaches, linear and circular, also show differences with respect to the spinning process step. This is due to changes in the supply chain. Precisely, the geographical location of the spinning process is different in the two case studies and therefore also the energy mix used in the spinning process is different. In the first case study, the spinning process is conducted in Turkey whereas in the second case study it takes place in Italy. Nonetheless, assuming spinning would be conducted only in Turkey, recycling provides the same trends in all environmental impact factors.

However, the recycled garment still contains 30% virgin cotton for quality reasons. The analysis clearly shows that the recycling process contributes at most 16% to the total life cycle impacts, while the rest mostly stems from the 30% virgin cotton production (see Fig. 4). Specifically, it can be seen that CR and MR processes contribute only 11% to total greenhouse gas emissions (see Fig. 4). Adding virgin cotton production and ginning to the recycling process results in 25% of total life cycle greenhouses gas emissions.

Therefore, it can be concluded that the recycling process has profound advantages compared to garment production from 100% virgin cotton.

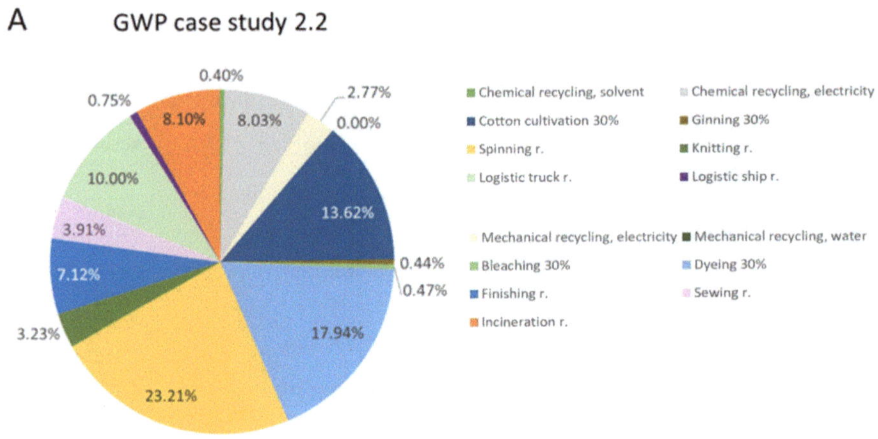

Fig. 4 Percentage impact from each process to the total of GWP impacts results

5 Discussion

The annual production of 326 thousand tons of post-industrial textile waste in the three North African countries examined, with the majority being landfilled, highlights the pressing necessity for implementing more circular waste management practices among EU suppliers in the Mediterranean region.

Notably, the analysis indicates that nearly half (48%) of this waste can be more aptly processed through new advanced technologies such as CR, rather than traditional MR methods. This observation not only underscores the shortcomings of existing recycling methods but also showcases the potential of cutting-edge technologies to revolutionize the industry's waste management strategies.

Egypt stands out in the study due to its large feedstock pool, which makes the country a candidate for investment in CR technologies that require a significant minimum efficient scale, and an investment size much higher than for traditional MR. A key risk associated with the investment is the recent volatility in the price of recycled PET (rPET), which has seen dramatic fluctuations over the past few years. The feasibility study pinpointed the price threshold at which the investment becomes profitable, achieving positive gross profit margins and internal rates of return (IRR) that surpass the weighted average cost of capital (WACC).

Conversely, the analysis of the business case for modern MR of cotton across the three countries underscores that mechanical technologies continue to play a crucial role in the strategy for textile waste valorisation. CR and MR technologies are more complementary than in competition in the valorisation of textile waste within the CE framework.

The study indicates that capital amortization represents the majority of production costs in contemporary recycling technologies, making up approximately half of the cost of producing recycled fibre. This suggests a need for financial models capable of supporting such investments, potentially through public-private partnerships, particularly in CR where the investment size is considerable.

Breaking the vicious cycle caused by the lack of recycling capacity, absent demand for waste recycling, and extremely low and ineffective post-industrial waste collection rates requires actions that complement the increase in recycling production capacity and the modernization of waste handling. The issue of low collection rates must be addressed at the level of the waste-originating companies, with training on the identification, collection, segregation, and management of textile waste.

Finally, the study emphasized the critical need to update the legal framework in the three countries, although the degree of urgency varies among them. This update is deemed essential to complement the introduction of modern technologies, initiating the transition towards a circular textile production system.

Considering also that the environmental sustainability issue is largely satisfied and respected, it is certainly desirable to adopt and develop the technology discussed.

6 Conclusion and Future Research Directions

In conclusion, the research conducted presents a compelling case for the adoption of both CR and modern MR technologies in North Africa's textile waste management strategies. By acknowledging the economic, technological, and environmental dynamics at play, these countries can pave the way for more sustainable practices that not only mitigate waste but also harness its value in a circular economy. Yet, the path forward requires careful consideration of investment risks, market volatility, and the broader economic context to ensure that these recycling ventures can achieve long-term sustainability and profitability.

In addition to technological advancements in the waste valorisation supply chain from a circular economy perspective, the challenges of updating the legal framework for waste management and training textile companies in proper waste management also require attention.

Further research is also needed to:

- Understand the dynamics of the informal section in the waste management markets with the aim to explore means and policy to upgrading the informal sector and promoting its integration. A significant portion of the recycling value chain still operates in the informal market, in all the three countries, limiting the overall efficiency and significantly reducing transparency and traceability in the value chain. Policy actions aimed at integrating the informal sector pose a real challenge for governments.
- Explore the potential of local circular business opportunities based on the industrial symbiosis model, especially regarding the end-uses of lower-quality recycled fibres in nonwoven applications in sectors such as agriculture, construction, filtering, furniture, and automotive.

- In addition to the deepening of the current LCA study with the use of additional indicators and/or integrating circular processes into the model, evaluate the reduction of local environmental impact that can be achieved at the country level with the development of a post-industrial textile waste value chain.

References

1. Division United Nation Statistics, "UN Comtrade." [Online]. Available: https://comtradeplus.un.org/
2. UNIDO, "SwitchMed, Circular Economy in Egypt." [Online]. Available: https://switchmed.eu/country-hub/egypt/
3. UNIDO, "SwitchMed, Circular Economy in Tunisia." [Online]. Available: https://switchmed.eu/country-hub/tunisia/
4. UNIDO, "SwitchMed, Circular Economy in Morocco." [Online]. Available: https://switchmed.eu/country-hub/morocoo/
5. Blumine s.r.l., "Toward an efficient and competitive circular textile industry - National Roadmap for minimizing and valorizing pre-consumption textile waste, Morocco," 2023. [Online]. Available: https://switchmed.eu/wp-content/uploads/2024/01/Morocco_EN-Toward-an-efficient-and-competitive-circular-textile-industry_low.pdf.
6. K. Saha, P. K. Dey, and V. Kumar, "A comprehensive review of circular economy research in the textile and clothing industry," *J Clean Prod*, vol. 444, p. 141252, Mar. 2024, doi: https://doi.org/10.1016/j.jclepro.2024.141252.
7. B. Khandelwal *et al.*, "Circular Strategies for a Garment's End-of-Life Moment," 2023. [Online]. Available: https://www.bcg.com/publications/2023/end-of-life-recycling-in-the-fashion-industry
8. Ellen MacArthur Foundation, "Fashion and the circular economy – deep dive." [Online]. Available: https://www.ellenmacarthurfoundation.org/fashion-and-the-circular-economy-deep-dive
9. M. Tripathi *et al.*, "Recent technologies for transforming textile waste into value-added products: A review," *Curr Res Biotechnol*, vol. 7, p. 100225, 2024, doi: https://doi.org/10.1016/j.crbiot.2024.100225.
10. B. Stubbe, S. Van Vrekhem, S. Huysman, R. G. Tilkin, I. De Schrijver, and M. Vanneste, "White Paper on Textile Fibre Recycling Technologies," *Sustainability*, vol. 16, no. 2, p. 618, 2024, [Online]. Available: https://doi.org/https://doi.org/10.3390/su16020618
11. S. Abrishami, A. Shirali, N. Sharples, G. E. Kartal, L. Macintyre, and O. Doustdar, "Textile Recycling and Recovery: An Eco-friendly Perspective on Textile and Garment Industries Challenges," *Textile Research Journal*, May 2024, doi: https://doi.org/10.1177/00405175241247806.
12. R. B. Baloyi, O. J. Gbadeyan, B. Sithole, and V. Chunilall, "Recent advances in recycling technologies for waste textile fabrics: a review," *Textile Research Journal*, Nov. 2023, doi: https://doi.org/10.1177/00405175231210239.
13. P. Luu, K. Daly, B. Croke, G. Sherwin, and B. Miñana, "Transitioning to a Circular System for Plastics," 2022. [Online]. Available: https://www.closedlooppartners.com/wp-content/uploads/2021/11/CLP_Transitioning-to-a-CIrcular-System-for-Plastics_3.pdf
14. Blumine, CSIL Milano, "Pre-feasibility study for the investment in a chemical recycling plant in Egypt.", Milano (2024). [Unpublished manuscript]
15. J. Fernando, "Discounted Cash Flow (DCF) Explained With Formula and Examples." [Online]. Available: https://www.investopedia.com/terms/d/dcf.asp#:~:text=How Does Discounted Cash Flow, value of the initial investment.
16. Council of the EU, "Green transition: Council gives its final approval to the ecodesign regulation," 2024. [Online]. Available: https://www.consilium.europa.eu/en/press/press-releases/2024/05/27/green-transition-council-gives-its-final-approval-to-the-ecodesign-regulation/

17. European Commission, "EU strategy for sustainable and circular textiles."
18. T. Duhoux, E. Maes, M. Hirschnitz-Garbers, K. Peeters, and et al., "Study on the technical, regulatory, economic and environmental effectiveness of textile fibres recycling – Final report," 2021. doi: https://doi.org/10.2873/828412.
19. K. Phan *et al.*, "Analysing the potential of the selective dissolution of elastane from mixed fiber textile waste," *Resour Conserv Recycl*, vol. 191, p. 106903, Apr. 2023, doi: https://doi.org/10.1016/j.resconrec.2023.106903.

Plastic Circularity as a Competitive Model: A Business Case in Toy Sector

Alessandro Fontana, Andrea Castrovinci Cercatore, Deborah Leone, Marco Pirotta, Marco Spaggiari, and Filippo Gallizia

Abstract The global toy industry is a major consumer of plastic, making it a critical sector for addressing plastic waste issues. This paper examines the potential of circular economy practices, particularly through the use of innovative materials such as Polymer XI, in reducing environmental impact and fostering sustainability in the toy sector. Polymer XI, developed by Geomagworld SA in collaboration with SUPSI, is a composite material made from 100% recycled polymer matrix with natural fillers like wood essences. This study investigates the technical, economic, and environmental viability of implementing circular solutions in the toy industry, highlighting the challenges and opportunities. The paper combines real business case analysis with laboratory results and life cycle assessment (LCA) to provide a comprehensive view of the possibilities and constraints in achieving circularity and sustainability in the toy industry through innovative material solutions.

Keywords Plastic circularity · Circular economy · Sustainable materials

1 Introduction

The global toy market was valued at approximately \$101.9 billion in 2022 and is projected to grow to an estimated \$135 billion by 2028 [1]. This industry is notably one of the most plastic-intensive sectors, using around 40 tons of plastic for every \$1 million in revenue [2]. It is important to notice that about 90% of all toys are made from plastic, and recent studies indicate that 80% of these toys are not recycled, ultimately ending up in landfills, incinerators, or the natural environment [3]. Addressing this critical issue necessitates a shift towards reduction, reuse, and recycling, fostering a circular economy, as explained in a key objective of the European

A. Fontana (✉) · A. C. Cercatore · D. Leone · M. Pirotta · M. Spaggiari
University of Applied Science of Southern Switzerland, Viganello, Switzerland
e-mail: alessandro.fontana@supsi.ch

F. Gallizia
Geomagword SA, Novazzano, Switzerland

A. Podleisek et al. (eds.), *Advances in Circular Economy: Bridging Research and Practice*, Springer Proceedings in Earth and Environmental Sciences,
https://doi.org/10.1007/978-3-032-21547-5_13

Green Deal for achieving climate neutrality by 2050 [4]. Additionally, the EU's plastic strategy aims to improve the life cycle of plastic products, from design and production to use and recycling, through the development of new polymer families such as biopolymers [5].

Bioplastics offer numerous environmental benefits, including reduced emissions and the absence of harmful chemicals found in traditional plastics. However, the adoption of bioplastics faces significant challenges, such as low market share, high production costs, limited applications, performance issues and potential environmental problems, such as a higher carbon footprint under certain conditions and complications in waste management [6].

Recycled plastics also present environmental benefits, despite their low global recycling rate of less than 10% [7]. Recycling can reduce raw material consumption and decrease the volume of waste destined for landfills, among other benefits [8]. However, current recycled-based polymers are priced between 4 and 7 €/kg, making them unaffordable for many products. This high cost hinders commercial penetration and negatively impacts manufacturers, especially as they face increasingly stringent regulations.

One promising approach to fostering a circular economy within the toy industry (and more) is the development and use of innovative materials like Polymer XI developed in collaboration with SUPSI by Geomagworld SA, a toy production company based in Novazzano (Switzerland). Polymer XI exemplifies a material that aligns with the principles of a Circular Economy. It consists of a 100% recycled polymer matrix enhanced by natural fillers, such as wood essences. This composite material, compliant with European Directive 2009/48/EC, is designed to create a new generation of environmentally sustainable products. These products are characterized by the minimal use of polymeric components, which are exclusively recycled, and wooden components sourced from production waste or forest maintenance. The use of highly sustainable recycled materials will adhere to international toy and food safety very strict regulations.

In addition, recent studies involving the complete Life Cycle Assessment (LCA) of wood production suggest that burning wood waste is a suboptimal solution compared to recycling it. Despite international support for biomass as an energy source, effectively reusing wood waste could significantly lower the overall carbon footprint [9].

Is it feasible to develop truly circular solutions that are also technically strong in the toy sector and, more in general, in plastic material applications? Beyond the sustainability claims, is such a model viable in terms of business profitability and cost-effectiveness?

This article will delve into these research questions through an analysis of a real business case of new Polymer XI developed, considering the applicability of the solution in terms of economics and sustainability. The paper analyses the case of XI polymer that replaced the previous material used by Geomagworld SA, ABS, which was taken as a baseline for a technical, economic and sustainable analysis. Apart from the formulation itself, which encompasses some novelties aimed at achieving technical and safety benchmarks, the whole technology has been developed based on a circular economy model and finally assessed against standard ABS environmental impact.

Everything from the raw materials' sourcing to the production process until conceivable end-of-life scenarios are considered and assessed to objectively and numerically determine the sustainability performance of Geomagworld's circular strategy. While details regarding the exact composition are omitted due to intellectual property policies, Sect. 3 reports ample laboratory results and an LCA analysis to demonstrate the technical and environmental benefits of the new compound. Section 4 is then presenting a Business Model (BM) analysis carried out that is meant to evaluate the economic feasibility of the project. Starting from the specific case of Geomagword SA and toy sector, the results obtained shows an interesting possibility to extend the use of the developed polymer in other industrial contexts.

2 State of the Art

The main plastics used in toys belong to a couple of polymeric families, namely polyolefins (PE and PP), and styrenics (ABS and PS) [10]. Due to the demanding technical requirements (e.g., strength and flexibility) and safety compliance to EN 71 standard (the reference one in this sector) [11], plastics used in toys are commonly derived from virgin feedstocks' polymers. In Geomagword SA toys are generally based on ABS and PP polymers. These materials offer additional advantages such as the low-cost, easy manageability (i.e., purchase and logistic), and easy coloring, which is an important aspect when it comes to kids' products [12]. The European Standard for safety of toys EN 71 sets out essential safety requirements that toys must meet to protect the health and safety of children. It consists of several parts, spanning from the presence of chemicals and small parts, or other specific hazards, to mechanical and flammability performance. In the framework of the current activity, two parts of the Standard are of main interest, and therefore addressed: Part 1 covering the mechanical and physical properties, and Part 3 related to the migration of certain elements.

In the context of sustainability and circular economy standards, bioplastics are increasingly seen as a target for material development. However, the simple use of bioplastics does not guarantee 100% sustainability or circularity. Factors such as raw material sourcing, production processes and end-of-life management, often summarised in a LCA, need to be considered [13]. Studies have shown that some bioplastics cannot be recycled, reused or composted and that most are made from edible materials, raising concerns about competition with food production [14, 15]. Second-generation biobased materials, such as lignocellulose residues or inedible agricultural waste, offer a promising alternative with a lower environmental impact [16–18].

Fossil fuel-based plastics have been optimized over decades, benefiting from extensive process improvements and subsidies, while bio-based plastics are still developing. Comparative LCAs are essential to identify improvement areas and understand the true environmental impacts of both materials [19]. Additives and specific processing techniques can increase energy consumption and emissions in bio-based polymer production, offsetting some benefits [20].

Recent research explores enhancing bio-based polymers with renewable source additives, but these also introduce new environmental considerations [21, 22]. Bioplastics face challenges like availability, cost, and mechanical performance, slowing their commercial acceptance [23]. As an alternative, recycled plastics are gaining traction, with companies implementing the 3R approach (Reuse, Reduce, Recycle). Proper recycling management can significantly reduce environmental impacts and enhance profitability [24].

Wood Plastic Composites (WPCs) are polymeric compounds known for decades, typically using wood fillers (e.g., fir, maple, beech, birch) and a polymeric matrix (mainly polyolefins). Chemical compatibility between these components, which have opposite polarities, is often enhanced by additives, though these should be avoided in certain sectors. WPCs are notable for incorporating recycled content, primarily used in construction as alternatives to solid wood for outdoor decking and fences [25]. The environmental benefits of WPCs include utilizing recycled materials and reducing waste. Recent studies focus on improving WPC properties with additives and treatments, though some enhancements may negatively impact environmental performance compared to fossil fuel materials [26, 27], even though recent studies, states that wood recycling has lower environmental impacts than wood burn [28]. It turns out that circularity can be industrially implemented by combining recycled polymers with wood scraps, reaching both sustainability and profitability goals.

In the toy sector, reduction and recycling schemes are still evolving. Geomagworld SA, a toy production company based in Novazzano (Switzerland), developed the new WPC formulation called XI that uses a recycled polymer matrix and wood particles from waste or recycling streams as filler, aligning with circular economy principles.

3 Scientific Model

The research questions are addressed through a description of the new XI polymer with an analysis of technical performances and a comparative LCA with the ABS component.

Material Technical Specification

The innovation of Polymer XI is distinguished by several key factors that set it apart from existing solutions. Firstly, it utilizes 100% recycled raw material. While similar claims may be made by other producers, Polymer XI stands out by offering certified proof of its recycled content, adhering to standards such as the Roundtable on Sustainable Biomaterials (RSB) [28]. Additionally, Polymer XI does not rely on synthetic compatibilizer additives. Instead, it includes less than 5% of a specific, widely available natural compatibilizer, selected after extensive testing, ensuring compatibility and performance without synthetic additives. A significant technological challenge overcome by SUPSI R&D team was achieving a stable and high-performance mixture with a high content of recycled wood. 33% by weight of recycled wood content has been successfully incorporated, with a goal of reaching 35% prior to commercialization. Initially, this was achieved with one type of wood, widely available in the EU and globally, but efforts are ongoing to extend this to other wood species. Moreover, these innovations would be meaningless without ensuring compliance with EN standards for toy materials. Through rigorous mechanical testing, Polymer XI has demonstrated its ability to meet the necessary safety and performance criteria, confirming its suitability for use in toys.

To develop a complete business model and bring the new polymer to market, it must also meet also technical performance requirements. Therefore, following are describes the work carried out in the development and testing of the Xi polymer.

Polymer XI is compliant to the EN 71-1 standards [11], specific for toys, which prescribes three types of mechanical resistance tests. Specifically, the Geomagword SA pentagonal base was employed for the EN 71-1 characterization trials, the same piece has been also used as a reference in the LCA study (in the following). The XI samples can sustain: (1) a free fall from 1 m height, (2) compression with 10 kg for 10 s, and (3) the horizontal impact with 1 kg weight free falling from 10 cm height, in a similar way compared to the standard polymers currently used by the company (Fig. 1).

These three tests are somehow qualitative characterizations, since the response is a pass/fail outcome. Therefore, the intrinsic properties of the XI material were also pre-evaluated with microstructural analysis to verify the homogeneity and compatibility of the wood charge within the polymeric matrix under shear stress and ISO mechanical characterization to have indications about possible weaknesses. ISO trials include tensile, impact and scratch resistance tests and density measurements.

The chemical incompatibility between polypropylene (apolar) and wood (polar) is known, usually solved by the addition of specific compatibilizers, e.g. maleic anhydride grafted polypropylene or MAPP [29]. Avoiding any EN-71 non-compliant

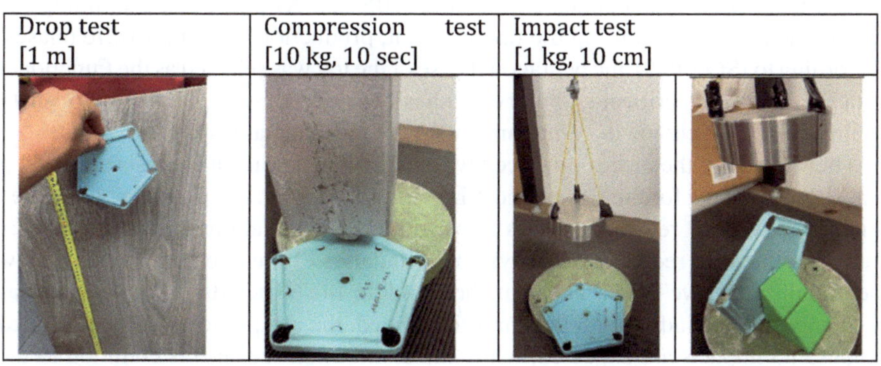

Fig. 1 Technical laboratory test

compatibilizer and process additives, specific mixing methods were adopted to obtain homogeneous and performing compounds such as the use of controlled oxidation of polypropylene and the addition of water steam. Eventually, a suitable natural compatibilizer was positively introduced to further enhance the chemical compatibility of the matrix and the filler.

Moreover, the maximization of wood essences is achieved by carefully tuning the extrusion parameters, such as mixing steps and screws profiles. The preparation of the compounds is carried out by means of a co-rotating two-screws extruder (Lesitritz ZSE 18HP-40D), at the IMP laboratory of SUPSI as well as in collaboration with SIP SA [30]. Industrialization will require a specific extruder. A patenting process of XI formulation is currently understudy, thus additional technical details are here omitted. Even though the individual raw materials (i.e., polymer, wood, and natural compatibilizer) are compliant to the limits set out by the EN 71-3 regulation, the measurement of ashes is implemented as a further assessment. A small amount of WPC compound is loaded into a muffle furnace and heated up to 600 °C for 30 min. The procedure is repeated twice, to confirm that the mass is not changing. The remaining ashes are weighed and related to the initial mass. It turns out that the residue is 0.26 wt%, consisting only in calcium carbonate ($CaCO_3$) and silica (SiO_2). The chemical composition is analyzed via Fourier Infrared Spectroscopy methodology (FTIR).

Life Cycle Assessment

To align with Geomagworld's goal of enhancing the environmental performance of its products, the LCA methodology was exploited. This approach assesses the environmental impacts of a new polymeric compound to determine its ecological advantages over the currently used material. The LCA is a widely utilized quantitative methodology defined by ISO 14040 for evaluating the environmental implications of a product, process, or service. The LCA framework consists of four steps: Goal and Scope Definition, Life Cycle Inventory, Impact Assessment, and Interpretation of Results. This comprehensive evaluation includes all stages of a product's life cycle, from extraction to disposal [31–33].

The LCA study was conducted for both ABS and XI materials, examining the environmental impacts during the raw material, production, and end-of-life phases. According to ISO 14040, the pentagonal base of the toy was selected as the Functional Unit (FU), as it has critical geometry and has been tested by the company to verify quality (Fig. 2), a choice in compliance with the mechanical test phase.

The weights of the pieces produced with the two materials in analysis are: 19.1 g for ABS and 19.6 g for the compound XI. The study is a "cradle-to-grave" LCA, i.e., it mainly covers all relevant process steps from raw material production to the final waste treatment or recycling. The system boundaries of the two compounds follow the "cradle to grave" approach, but the "middle life", including distribution has been currently excluded from the analysis. Indeed, the use phase has been excluded because no impacts are expected, while the distribution scenario is common to the two materials and therefore omitted. For ABS the analysis begins with the extraction of raw materials and ends with the incineration of the product. For compound XI the

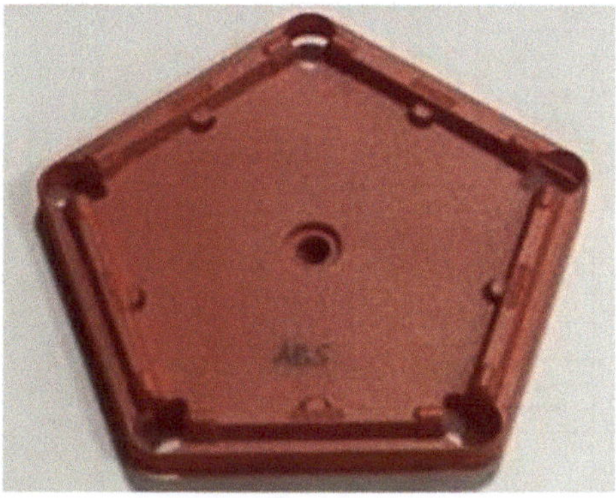

Fig. 2 Pentagonal base toy used as functional unit in the LCA study

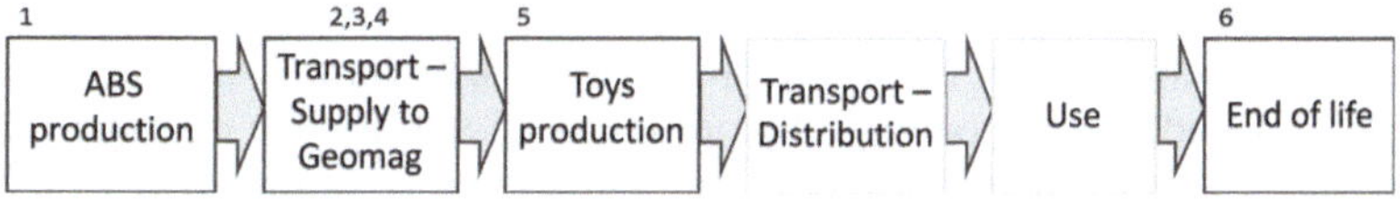

Fig. 3 ABS LCA process

analysis starts with the recycling of polypropylene and ends with the incineration. The transport between the process steps is included in both cases. Finally, it was decided to use the cut-off approach to manage the end-of-life impact. This decision was driven by the fact that the company uses recycled materials and scraps. Furthermore, there is very little control over the end-of-life phase, so it has been decided to model the incineration phase for both materials as a worst-case scenario.

Simplified process flow diagrams summarizing the production of the two compounds, including system boundaries, are presented in Fig. 3.

The first process in the ABS life cycle is the production of the material. The process includes all processes from raw material extraction until delivery at plant. Processes 2,3,4 include the transport of the material from the supplier sited in Asia to Geomagword SA. Process number 2 is the transportation by lorry in Asia to the port. Process number 3 is the transportation by ship from Asia to Europe and process number 4 is the transportation by lorry within Europe, from the port to Geomagword SA. The process number 5, "toy production", is the injection moulding of the pentagonal base. The End-of-life, process number 6, is the incineration of the injected part through a municipal incinerator.

The XI process flow diagram shows the processes considered in Fig. 4. Compared with ABS, a higher and different number of processes are listed due to the various processes required for the new polymer. Also, for this reason, a comparative LCA analysis is essential in order to be able to certify the actual reduction in environmental impacts despite the increase in number of processes involved.

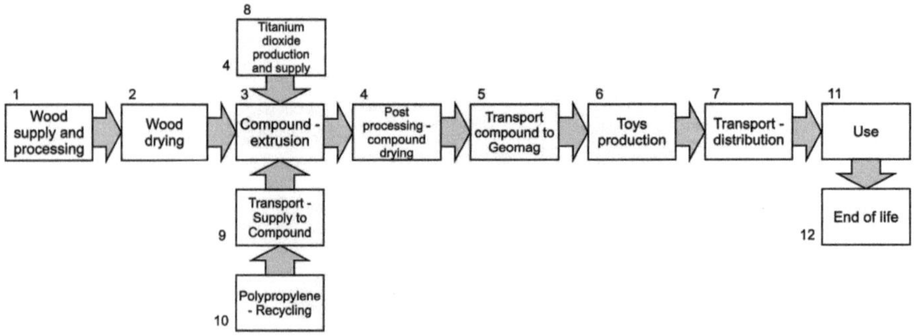

Fig. 4 Xi LCA process

The first process in the compound XI life cycle is "wood supplying and preparing". This process includes the transport from the wood source to the chopping facility (50 km) and the chopping of the wood into chips in the sawmill. Once the wood flour has been chopped, it must be dried before it can be fed into the extruder (process 2). The other steps of the life cycle are the same as those of the ABS seen above, except for the production of titanium dioxide (process 4). Titanium dioxide's production includes the activities that produce titanium dioxide and the supply of it to the consumer of the product. The End-of-life, process number 10, is the incineration of the injected part through a municipal incinerator.

The majority of Life Cycle Impact Assessment (LCIA) data used in the study were collected from the Ecoinvent database [34] (version 3.8) and from the Environmental Footprint database [35], even though most of them have been modified to better represent the specificities of the processes involved. For instance, specific energy consumptions and electricity mixes has been collected from the data retrieved by the various operators involved for the compound preparation (i.e. SUPSI and SIP) and transformation (Geomagword SA). In this LCA analysis, the Global Warming Potential (GWP) indicator has been calculated as representative of the environmental aspects to compare the two solutions. GWP is one of the main indicators used to compare a conventional plastic material with a bio-based one as described in the literature [36], and secondly because it is of great interest for company reporting. This represents a key starting point therefore for the company, to understand the capabilities and environmental improvements of the new Xi material. Subsequently, LCA will also be extended to other selected indicators. The GWP is calculated in carbon dioxide equivalents (CO_2-Eq.), meaning that the greenhouse potential of an emission is given in relation to CO_2. The results are show in Table 1 and Fig. 5 GWP result comparison describe the hotpots divided by process considered.

The new XI polymer has an environmental impact reduced by almost 50% when considering the GWP indicator. This significant improvement is largely attributed to the different raw material production processes, with ABS raw material production accounting for 60% of its total environmental impact, as show in Fig. 6. In the case of XI, the phase with the highest GWP impact is incineration, which contributes

Table 1 LCA result

Material	GWP (kg CO_2 Eq)
ABS	0.149
Xi	0.077

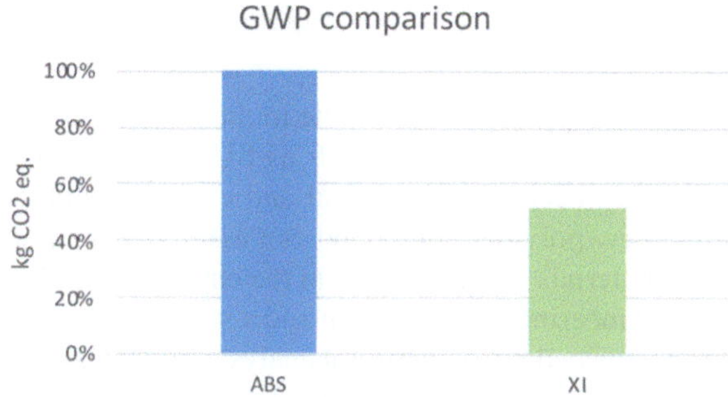

Fig. 5 GWP result comparison

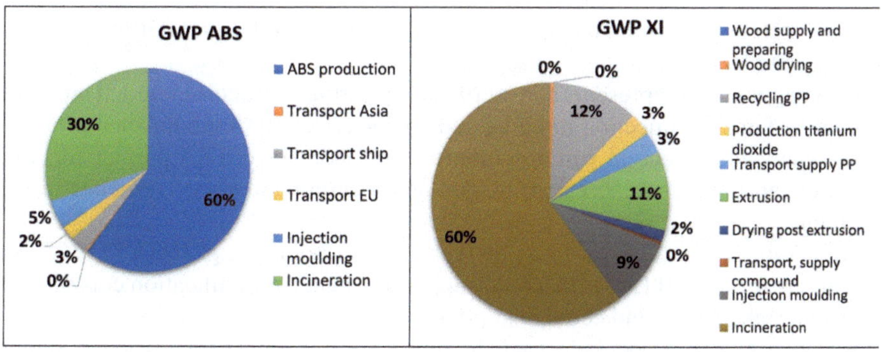

Fig. 6 GWP hotspot of ABS and Xi

60% of the total GWP impact. The second phase with the highest impact is injection moulding, which accounts for 12% of the total GWP impact. The extrusion process has a similar impact to injection moulding, contributing 11% of the total GWP impact.

4 Business Model

To answer the research questions presented above, some business considerations are provided using the strategic Business Model Canvas tool. Initially proposed by Osterwalder, the Business Model Canvas (BMC) is a strategic management template used for developing new business models and documenting existing ones. It describes, with a visual chart, a company's value proposition, infrastructure, customers, and finances, assisting companies to align their activities by illustrating potential trade-offs [37]. In the following the fulfilment of the BMC showing the evaluations carried out for each part of the chart: Key partnership; Key activities; Key resources; Value proposition; Customer relationship; Channels; Customer segments; Cost structure; Revenue streams.

Key Partnership

Several key partners are expected to contribute to the technical and business feasibility of the project. Research institutes such as SUPSI contributed to the material development and its mechanical, physical and environmental characterization to guarantee that the new polymer compound can assure the same (or even better) performances of the materials currently used in the envisaged applications. With an excellent industrial engineering research team and a state-of-the art polymer laboratory, SUPSI indeed performed microstructural analysis, shore test (ISO 868), density measurement (IAO 1183-1). Conversely, traction (ISO 527-1/2) and resilience tests (ISO 179) was carried out by another stakeholder, PROPLAST, collaborating to the future technical development of the compound to reach the 35% of wood content to assure the final mechanical EN 71-1 compliance. Again, in terms of R&D activities, research institutes and universities can support the ecodesign of the material, as already done by SUPSI.

Geomagword SA represents here a crucial key partner to be considered. It provides its own systems for injection molding and characterization of the finished product. Its commercial connection with toy industry via Geomagword SA partnership will guarantee an immediate access to market and a certain stream of revenues in the beginning.

An additional stakeholder class are certification bodies (e.g. INTERTEK [38], EPD International AB [39]) that could support the material certification considering for instance the safety and ecological points of view.

Eventually, also wood industrial districts or, more in general, wooden scraps producers (e.g. wood milling or furniture firms) must be considered since they are low-cost providers of raw material for Xi. Finally, the collaboration with national or regional governments is crucial for the production site settlement and the access of possible incentives. As an example of possible stakeholder clusters to be considered, the Italian region of Trento and Piemonte, the Switzerland Canton Ticino and Spain Southwest area are united by: (1) the presence of strong wood production districts (fed by forests in the case of Trento and Ticino, by almond shells for Spain); (2) widespread industrial clusters such as plastic districts in the case of Piemonte and Spain, toy district in the case of Alicante (south Spain), processors potentially interested in sustainable innovation in the case of Switzerland where it is easy to find skilled human resources, infrastructures and production capabilities; (3) strategic location within important industrial areas; iv. availability of provincial economic incentives.

Key Activities

The Xi BM is not only meant to envisage the commercialization of the material, but to provide a series of services that could technically support the final customer in the better application of the compound for the as-is material substitution. Clients' needs are addressed thanks to the compound customization in terms of composition, characteristics and colors. To assure market requirements R&D activities are crucial to continuously improve the material characteristics and to assure its environmental and circular benefits. The preparation of documentation and data is also fundamental to helps fuel in the customer the motivation to adopt Xi as an important strategic sustainability move. Eventually, the definition of the industrial production site and its positioning is also key. To better address CE principles, the production of Xi must be set near the production of wood wastes and in region that can offer economic incentives, technical and human expertise in plastic transformation.

Key Resources

Raw materials are surely fundamental resources for the Xi project. Recycled plastic must be available from carefully selected suppliers, able to provide raw material which is EN-71 compliant. The bio-based part (wood) must come from industrial waste or forest maintenance origin. The wood component can be retrieved by a very high number of producers across many EU regions. This will allow for more decentralized production, a significant use of local resources and a vast reduction in transport of raw materials. Small to medium production units could be envisaged, in the framework of an industrial scalability of the project, which are located where the demand of polymeric material meets the availability of the wood waste. Following this approach, the circular economy model could be further developed. As detailed in the Cost structure section of this BM canvas, a typical production unit will be constituted by: a wood flour mill; two extrusion lines for Wood Polymer Compound; one injection molding machine and a samples mold; one laboratory extruder.

Value Proposition

Based on extensive research and experimental activity, Polymer XI has been developed as an a highly innovative material to replace both traditional plastic materials and recycled plastics. The new material uses a low amount of recycled polymeric matrix (55 wt%) as a binder of natural materials of industrial waste or forest maintenance origin (up to 35 wt%). In addition, the XI compound, should have a very low cost and therefore allow the marketing of highly innovative products without asking the market to pay a surcharge that is usually associated with ecological or bio-based products. Moreover, the use of recycled and renewable materials will effectively lower the environmental impact of any product made of such polymer. The sum of these factors, linked to the possible polymer colorability as close as possible to common polymers (ABS and PP) represents an almost immediate opportunity for innovation of any existing product.

Although the polymer-wood composite material is widely used in various sectors, the toy sector is still lacking a serious diffusion. Polymer Xi is a real innovation for its mechanical performance, natural additives compliant to safety regulations for toys and 100% recycled materials content. Thanks to the wood content, it is foreseeable a sale price up to 50% lower than competitors' solutions, thus addressing the business challenge to keep an affordable price.

Moreover, polymer Xi will have a positive impact on industry, society and the environment: it will allow an increased utilization of waste/recycled wood material instead of plastic, even if recycled, reducing the overall impact of any product realized with such material. Polymer Xi could also be a starting point for a new generation of eco-friendly polymers with different applications being a highly innovative EU made product that can create a new segment leveraging the advantages of superior technology and short/new value chain.

The competitive benefits focus on the following Unique Selling Points (USPs): (1) made in the EU with a focus on the sustainability of the industrial process; (2) use of 100% recycled materials (polymers and wood) and guaranteed certificate of origin for the wood component; (3) price up to 50% lower than competition; (4) availability of a high-level R&D Department to support customer's needs.

Customer Relationship

As already stated in the Key activities section, several strategic levers addressing the customer needs and guaranteeing a valuable relationship are identified. Clients are provided with documentation such as technical data sheets, LCA results, training and education are required to proof the innovation of Polymer Xi compared to oil-based solutions, standard recycled polymers and bioplastics. Most plastic transformers are still tied to standard oil-based products and increased awareness on the sustainability topic is requested. In addition, clients must be assisted during the implementation of the new material within their proven production process so that the new compound cold be seamless integrated in the current production to substitute oil-based or biobased polymers without any qualitative problem. Eventually, the affiliation with customer is also guided by the customization possibilities offered by the new compound since it is possible to modify the base Polymer Xi blend to meet specific customer's requirements in terms of quality or process or product.

Channels

The first target of Xi commercialization will be industrial players (Business to business—B2B) characterized by a significant use of plastics, mainly in the Toy Industry but also from different industries (packaging, medical, home decoy and others) who could either directly purchase or manufacture themselves Polymer Xi under a licensing contract agreement. Commercialization will be pushed through a commercial dissemination in specialized trade fairs, such as, K Trade Fair and Plast, and with a dedicated sales team.

Customer Segments

Pre-commercial agreements set with Geomagword SA will produce the first revenues in the first 2 years. Moreover, some interest has been also collected from other industrial sectors: the business model will target companies and compounders which uses polymers for toys and/or other products. Concerning toys, the market targeted by this innovation is emerging there is a growing demand, and few offerings are available. The degree of innovation about the development of sustainable production technologies is rather low. This is mainly due to the dependence on almost exclusively Asian external productions, which are not willing to provide innovative products from the point of view of materials. The market, however, is increasingly sensitive to the issue of environmental sustainability and does not find performant products at a competitive price.

The interest for this polymer will not only arise from the toy industry but also from other industries which have intensive use of plastic since fossil-based plastic production is still growing, reaching around 396 million metric tonnes in 2022 [40]. For example, the packaging industry has been recently making efforts the reduce plastic consumption and could be very benefited by Project XI. Recycled plastic is surely spreading (recycled content in packaging having grown by '22% year-on-year [40], but recycling plastic is a suboptimal solution and Polymer XI will effectively replace a significant share of recycled plastic with natural, cheap, widely available, waste wood.

Cost Structure

The present formulations of Polymer XI is the results of direct R&D investments made by SUPSI (backed by Innosuisse, grant project ID 47891.1) and a direct contribution by Geomagword SA with more than €100 k. Some other financial resources are being spent for IP activities. Finalizing the formulation and the consequent industrialization will require further investments. The project indeed will need Capex investments for more than €500 k that are including the following elements:

- A wood flour mill with the production capacity approximately 450 kg per hour;
- Two specific extrusion lines for Wood Polymer Compound with unit production capacity of about 350 kg per hour. The second line installation is planned during the third year of activity;
- One 80 tons injection molding machine for testing, and a samples mold;
- One laboratory extruder to support R&D and customer care activities, a test room to carry out quality tests.

Other productive investment and evaluations in the next 5 years include:

- Human resources: starting from 7 full time equivalent (FTE) employees for the first year, reaching 18 FTE in the fifth year, including a CEO, laboratory engineers, employees on extruder and mill, commercial and assistant manager and warehouse employees.
- Increasing utilization of the production plant is expected in the first 5 years from 220 up to 350 days per year normally on 2 shifts. This plant structure allows production of up to more than 5000 tons of compound per year. The availability of 2 extruders, in addition to diluting the investment in the first 3 years, allows for more flexible production suitable for supplying customized product.

Revenue Streams The current toy market, which reached a market size of $130Bn in 2020 [41] has been steadily growing in recent years and a projected CAGR of 7.3% for 2021–2028. Considering that plastic toys still account for 2 of every 3 toys sold, the end-customer sales of plastic toys represent about $85Bn. Based on Geomagword SA experience, the cost of plastic as raw material represents approximately 1% of sale price. This means that the expenditure in plastic within the toy industry equals to €850 m, out of which 30% could be replaced by the new polymer, leading to an addressable market of €250 m. Bearing in mind the alliance with Geomagword SA, the first target market would thus be construction toys, in which Geomagword SA is specialized with €50 m revenues, and would be the Serviceable and Obtainable Market (SOM) in the short term.

The BM plans to incorporate an independent company that will carry out the industrialization and the commercialization of Polymer Xi and other possible future developments. In 5 years, Polymer XI is expected to generate revenues in the order of €15 million, the predicted SOM. Moreover, an Earnings Before Interests Taxes Depreciation and Amortization (EBITDA) conservatively in the range of 20–25% has been foreseen. Furthermore, the innovative formulation and the envisaged modular production will allow a decentralized production which could be replicated in other EU countries. Even if it's too preliminary at this stage, it is not possible to exclude the development of a new generation of polymers following the XI's success. Indeed, it's worth noting that, other possible applications, such as packaging, constitute a multi-trillion € market [42].

5 Conclusion

This paper presents the case of XI, an innovative polymer developed by Geomagword SA in collaboration with SUPSI. The compound has been conceived and developed to address the Circular Economy principles, being composed by a recycled plastic matrix that is compatibilized through bio-based substances with wood flour obtained from scraps and wastes. The main objective of this publication is to demonstrate the feasibility of the conceived solution considering both technical and economic evaluations. The polymer must guarantee performances similar of those showed by oil-based or recycled plastics currently used in the toys sector so that it could represent a valuable alternative for the traditional material substitution. This is done considering also the economic aspects of the compound that must assure an appropriate price to address mass production markets. In this perspective, a business model analysis is presented to demonstrate the financial viability of the project and to structure a possible related sustainable business. Moreover, a LCA analysis has been also performed to verify and assure that the innovative compound is showing better environmental performances. The science-based methodology allows the calculation of the material GWP and demonstrates that the polymer has about the halves of the carbon footprint in comparison with oil-based plastic usually employed in toys construction. The promising results obtained demonstrate that the circular solution developed can assure technical advantages, profitability and lower environmental burdens, suggesting a possible extension of its use in other industrial sectors besides the toys.

As next steps, Geomagword SA is working on refining the formulation, with the objective to reach the 35% of wood content, a value that would secure better mechanical properties, an improved use of bio-based recovered material form waste together with a lower plastic content (even though recycled). A better industrialization of the polymer is also ongoing to refine the extrusion process and to set the actual industrial production of the compound to bring Polymer XI to the mass production. Intellectual Property (IP) and commercialization activities are also ongoing to protect the material development and to launch the polymer on the market that will be realized starting from existing partner (as Geomagword SA) but with the aim to build an extensive network of toy and packaging producers. Finally, the sustainability analysis will be extended on other environmental indicators to avoid problem shifting that could be generated improving the performances in certain areas (e.g. GWP) to the detriment of other environmental compartments. A comparison amongst XI and another recycled polymers, such as the one currently used by Geomagword SA constituted by 100% recycled PP and a mineral charge, will be also investigated with the aim to verify that Polymer XI cold be competitive also with materials that have already been identified as more sustainable compared to oil-based plastics.

References

1. CEPSA (2023). Eco-friendly Toys: Discarding Microplastics. Planet Energy. Available at: https://www.cepsa.com/en/planet-energy/2030-goals/eco-friendly-toys-discarding-microplastics
2. Raynaud, J., Richens, J., & Russell, A. (2014). The Business Case for Measuring, Managing and Disclosing Plastic Use in the Consumer Goods Industry.
3. Robertson-Fall, T. (2021). Creating a Circular Economy for Toys: Solutions to Design Out Waste and Pollution. Toy Industry Journal.
4. European Commission (2022). European Green Deal: Putting an End to Wasteful Packaging, Boosting Reuse and Recycling.
5. European Commission (2018). EU Plastics Strategy. Available at: https://environment.ec.europa.eu/strategy/plastics-strategy_en
6. Costa, A., Encarnação, T., Tavares, R., Todo Bom, T., & Mateus, A. (2023). Bioplastics: innovation for green transition. Polymers, 15(3), 517.
7. SpecialChem. Bioplastics and Circular Economy: A Pathway to Sustainability Goals. (2023)
8. Mwanza, B.G. (2021). Introduction to Recycling. In: Parameswaranpillai, J., Mavinkere Rangappa, S., Gulihonnehalli Rajkumar, A., Siengchin, S. (eds) Recent Developments in Plastic Recycling. Composites Science and Technology. Springer, Singapore. https://doi.org/10.1007/978-981-16-3627-1_1
9. Geyer, R., Jambeck, J. R., & Law, K. L. (2017). Production, Use, and Fate of All Plastics Ever Made. Science Advances, 3(7), e1700782. https://doi.org/10.1126/sciadv.1700782
10. British Plastics Federation. Biodegradable Polymers In the Toy Sector Available at: https://www.bpf.co.uk/article/biodegradable-polymers-in-the-toy-sector-364.aspx
11. European Committee for Standardization (2019). EN 71: Safety of Toys – Part 1 & Part 3.
12. Sanchez, S. (2021). Changing the World Toy by Toy. Toy Sustainability Review, 4(1), 21–30.

13. Rosenboom, JG., Langer, R. & Traverso, G. Bioplastics for a circular economy. Nat Rev Mater 7, 117–137 (2022). https://doi.org/10.1038/s41578-021-00407-8
14. Pascoe Ortiz, S. (2023). Are bioplastics the solution to the plastic pollution problem?. Plos Biology, 21(3), e3002045.
15. Bishop, G., Styles, D., & Lens, P. N. (2021). Environmental performance comparison of bioplastics and petrochemical plastics: A review of life cycle assessment (LCA) methodological decisions. Resources, Conservation and Recycling, 168, 105451.
16. Von Braun, J. (2018). Bioeconomy. The global trend and its implications for sustainability and food security. Glob. Food Sec, 19, 81–83.
17. Lewandowski, I. (2015). Securing a sustainable biomass supply in a growing bioeconomy. Glob. Food Sec. 2015, 6, 34–42.
18. Hjuler, S. V., & Hansen, S. B. (2018). LCA of Biofuels and Biomaterials. In Life Cycle Assessment (pp. 755–782). Springer, Cham. https://doi.org/10.1007/978-3-319-56475-3_33
19. European Bioplastics (2022). Using LCA to Compare Bio-based and Fossil-based Plastics is Not That Simple. Available at: https://www.european-bioplastics.org/using-lca-to-compare-bio-based-and-fossil-based-plastics-is-not-that-simple
20. Ita-Nagy, D., Vázquez-Rowe, I., Kahhat, R., Chinga-Carrasco, G., & Quispe, I. (2020). Reviewing environmental life cycle impacts of biobased polymers: current trends and methodological challenges. The International Journal of Life Cycle Assessment, 25, 2169–2189.
21. Marturano, V., Marotta, A., Salazar, S. A., Ambrogi, V., & Cerruti, P. (2023). Recent advances in bio-based functional additives for polymers. Progress in Materials Science, 101186, 139,
22. Walker, S., & Rothman, R. (2020). Life cycle assessment of bio-based and fossil-based plastic: A review. Journal of Cleaner Production, 261, 121158.
23. Sidek, S., Izathul, I., et al. (2019). Current Development on Bioplastics and Its Future Prospects: An Introductory Review. I-Tech Magazine, 1, 3–8.
24. Raynaud, Julie. (2014). The Business Case for Measuring, Managing and Disclosing Plastic Use in the Consumer Goods Industry.
25. Xu, K., Du, G., & Wang, S. (2021). Wood Plastic Composites: Their Properties and Applications. In Composite Materials: Advances in Science and Technology
26. Schwarzkopf, M. J., & Burnard, M. D. (2016). Wood-plastic composites—Performance and environmental impacts. Environmental impacts of traditional and innovative forest-based bioproducts, 19-43.
27. Morris, J. (2017). Recycle, bury, or burn wood waste biomass?: LCA answer depends on carbon accounting, emissions controls, displaced fuels, and impact costs. Journal of Industrial Ecology, 21(4), 844-856.
28. Roundtable on Sustainable Biomaterials (RSB) (2023). Certification and Standards for Bio-based Materials. Available at: https://rsb.org/
29. Kim, H. S., Lee, B. H., Choi, S. W., Kim, S., & Kim, H. J. (2007). The effect of types of maleic anhydride-grafted polypropylene (MAPP) on the interfacial adhesion properties of bio-flour-filled polypropylene composites, 1473, 1482, 38.
30. SIP SA (2023). Industrial Processes and Technologies for WPC Extrusion. Available at: https://www.sipsa.ch
31. ISO 14040 ISO (2006). ISO 14040: Environmental Management – Life Cycle Assessment – Principles and Framework. International Organization for Standardization.
32. Tamburini, E., Costa, S., Summa, D., Battistella, L., Fano, E. A., & Castaldelli, G. (2021). Plastic (PET) vs bioplastic (PLA) or refillable aluminium bottles–What is the most sustainable choice for drinking water? A life-cycle (LCA) analysis. Environmental research, 196, 110974.
33. Wolfson, A., Dominguez-Ramos, A., & Irabien, A. (2019). From goods to services: the life cycle assessment perspective. Journal of Service Science Research, 11, 17–45.
34. Ecoinvent (2022). Ecoinvent Database, Version 3.8. Available at: https://ecoinvent.org/database
35. European Commission (2022). Environmental Footprint Database. Available at: https://eplca.jrc.ec.europa.eu/EnvironmentalFootprint.html
36. Hill, C. (2018). Biopolymers (Bio-based Plastics): An Overview. JCH Industrial Ecology Ltd. Available at: https://www.bioplasticsmagazine.com/en/journal/archive/2018

37. Osterwalder, A., & Pigneur, Y. (2010). Business Model Generation: A Handbook for Visionaries, Game Changers, and Challengers. Wiley.
38. Intertek (2023). Toy Safety Certification Services. Available at: https://www.intertek.com/toys-childrens-products/certification
39. EPD International AB (2023). Environmental Product Declarations.
40. Statista (2023). Global Plastics Production 1950–2022. Available at: https://www.statista.com/statistics/282732/global-production-of-plastics-since-1950
41. Fortune Business Insights (2023). Global Toys Market Forecast 2021–2028. Available at: https://bit.ly/33NHMwe
42. Market Research Future (2023). Packaging Materials Market: Trends and Forecast. Available at: https://bit.ly/3g6bJKY

CON: A Case Study on a Structural System Utilizing Reused Concrete Slabs

Adrian Kiesel, Guido Brandi, Patric Fischli-Boson, and Andri Gerber

Abstract Concrete production is responsible for approximately 8% of global CO_2 emissions, largely due to the energy-intensive process of cement manufacturing. One effective strategy to reduce these emissions is the reuse of concrete elements from buildings slated for demolition. Unlike recycling, which requires energy-intensive processes to break materials down to their raw state, reuse preserves the original form, mechanical properties, and chemical integrity of components. This paper presents a proof-of-concept prototype that explores the reuse of concrete slabs within a 1:1 mockup for a multi-story structural system. Emphasis is put on the ability of the system to allow the integration of slabs with varying thicknesses as well as testing the performance of two construction design iterations. This adaptability is achieved through the development of new Design for Disassembly (DfD) connection methods, enabling the incorporation of different slab types into the design. A life cycle assessment (LCA) reveals the performance of the system which reduces global warming potential (GWP) by 62.6% compared to a traditional construction using concrete and by 67.9% when compared to a traditional construction DfD systems made of concrete. Altogether, the findings of this paper highlight the potential for reused concrete slabs to meet structural requirements while substantially reducing embodied carbon. This research contributes to the growing body of work on circular construction and advocates for broader adoption of material reuse strategies to reduce the environmental impact of the construction industry.

Keywords Reuse · Structure · Hybrid system · Steel · Concrete · Design for disassembly · Sustainability · Future of construction · Reuse concrete

A. Kiesel (✉) · G. Brandi · A. Gerber
ZHAW School of Architecture, Design and Civil Engineering, Institute of Constructive Design, Winterthur, Switzerland
e-mail: kied@zhaw.ch

P. Fischli-Boson
ZHAW School of Architecture, Design and Civil Engineering, Institute for Building Technologies and Processes, Winterthur, Switzerland

A. Podleisek et al. (eds.), *Advances in Circular Economy: Bridging Research and Practice*, Springer Proceedings in Earth and Environmental Sciences,
https://doi.org/10.1007/978-3-032-21547-5_14

1 Introduction

1.1 Context

Concrete production is responsible for approximately 8% of global CO_2 emissions, primarily due to the energy intensive process of producing cement [1–5]. Concrete is the most used building material globally, with estimations suggesting that approximately 30 billion tons are used annually.

Efforts to reduce emissions in the construction sector have recently focused on operational emissions, which occur during the building's use phase and consist of energy consumption for heating, cooling, and lighting [6].

To reduce a building's CO_2 footprint, it is essential to not only optimise operational emissions but also to account for embodied emissions, which arise during the extraction, manufacturing, transportation, and assembly of a building and its materials. It is reported that these emissions represent a substantial portion of a building's overall carbon footprint [7]. Reducing embodied emissions is crucial for minimizing the environmental footprint of the construction industry. One effective and immediate strategy for reducing embodied emissions in construction is the reuse of materials, particularly structural elements such as concrete slabs and beams salvaged from buildings slated for demolition. By preserving the physical integrity of these components, the need for producing new materials is avoided, eliminating emissions associated with the production, transportation, and processing of new concrete. This approach not only extends the lifecycle of construction materials but also reduces resource extraction and the material waste [8–10].

1.1.1 Circularity within the Building Industry

Circular construction is based on the principles of the circular economy. The goal is to reduce waste, to optimize and to extending the lifespan of resources throughout the entire use-cycle of materials. This goal is achieved by keeping materials in use for as long as possible, thereby reducing the need for new materials. The primary strategies of circularity are defined within the 10R hierarchy, which include refuse, reduce, redesign, reuse, repair, refurbish, remanufacture, repurpose, recycle, recover. Originally defined from waste management operations, these strategies aim to close the loop within different value hierarches. By operating within the various hierarchies, waste can be minimized, resource efficiency improved, and environmental impact reduced [11, 12].

The reuse of structural concrete components aims to operate within the fourth hierarchy, 'reuse,' elevating the process from the traditionally applied ninth hierarchy, 'recycling.' Unlike recycling, which typically involves breaking down concrete into aggregate and reprocessing only the aggregate, reuse preserves the structural integrity of concrete elements. This approach minimizes the energy-intensive processes involved in both producing new elements and recycling.

This approach integrates salvaged concrete slabs, walls, columns, and beams directly into new construction projects, enabling their continued use. Implementing concrete reuse presents several challenges, including the need for precise planning, careful deconstruction, testing, if necessary, adaptation, and transportation of the components to the new construction context. Unlike conventional building demolitions, where non-load-bearing materials are stripped of the building before the structure is being destroyed, harvesting structural elements requires controlled deconstruction to recover the desired pieces, to prevent damage and to ensure the structural integrity of the components for future applications. Once the elements are successfully recovered, they can be inspected, tested for damage, and assessed for their reuseability. It is important to note that these steps contribute to the overall environmental impact of the recovered components, which should be closely monitored throughout the entire process. If a component meets the designers' specifications, it can be transported directly to the new building site for immediate reuse or taken to a workshop for modifications to meet the requirements of the new structure. To facilitate future life cycles, design strategies such as Design for Disassembly (DfD) and Design for Manufacture and Assembly (DfMA) offer significant potential to enable more efficient reuse of components. DfD emphasizes creating buildings that can be easily dismantled, allowing for efficient recovery of components, while DfMA focuses on designing elements that can be easily assembled and disassembled. Although the initial costs and complexity of these strategies may be higher compared to conventional approaches, life cycle assessment (LCA) studies demonstrate that the overall ecological impact of reusing concrete components is significantly lower than producing new materials [13].

The practice of reusing construction materials is not a new concept. Before industrialization, material costs were often higher than labor costs associated with modifying and repurposing elements. This led to construction systems prioritizing reuse or material efficiency, which, although labor-intensive, was economically preferable. Repurposing building materials through multiple reuse cycles was once the standard practice in the construction industry. Good examples are elements made from stone and wood, which were frequently reused due to their high value and durability, making them ideal for incorporation into new structures. However, with the advance of industrialization and the optimization of production processes, the focus shifted toward mass production, which reduced material costs drastically. Today labor costs have become the primary factor driving up overall construction expenses. As a result, time-intensive processes such as material reuse have declined in favor of newly produced materials. The labor-intensive nature of disassembly and reassembly has made reuse economically unviable compared to the cost-effective production of new materials [14].

In recent years, growing concerns over climate change and the depletion of natural resources have prompted renewed interest in reusing structural materials. As the environmental impacts of producing new building materials have become clearer, the reuse of materials takes on traction again as a sustainable approach for construction. However, the reuse of large structural components presents several challenges, particularly regarding logistics, structural integrity, and adaptability to the design of the new building and its requirements [15].

Despite these challenges, recent case studies have shown that reused elements can successfully meet both structural and regulatory standards while significantly reducing the embodied carbon footprint of buildings. Full-scale projects, like the Kopfbau Halle 118 build by Baubüro in situ in Winterthur utilized various reused materials within the design of the building and were therefore able to lower CO_2 emissions by 59% [16–18]. To further foster understanding reuse process, it is essential to promote and analyze case studies that utilize reused materials. As this approach continues to evolve, it offers a immediate solution for reducing the environmental impact of the construction industry.

1.2 Objective

The primary objective of this paper is to assess the feasibility, benefits, and challenges of reusing concrete elements, particularly structural slabs, in new construction projects. Central to this research is the construction of a prototype system in a 1:1 mockup, which serves as a test platform to evaluate both the processes involved and the performance of the construction system. This mockup enables the practical examination of controlled deconstruction, as well as the development and testing of new DfD connection methods, which are designed and implemented in the reassembly of the components. The mockup also facilitates the comparison of two construction variants, investigating different design strategies to optimize the integration of slabs with varying thicknesses.

To assess the environmental performance of the construction, this paper explores the global warming potential (GWP) of concrete reuse by performing a comparative LCA of the mockup and comparing it to traditional concrete construction methods and DfD systems using new materials.

This study contributes to the growing body of research on circular construction, particularly focusing on concrete, one of the most environmentally intensive materials in the building industry. By demonstrating the viability of reclaimed concrete elements through mockup testing, this research advocates for the broader adoption of material reuse strategies, offering practical insights into how reused materials can be systematically integrated into modern construction to reduce the industry's overall GWP.

2 Methods and Materials

2.1 Methodology

This paper employs a systematic approach to the reuse of concrete elements within the framework of circular construction. It defines the procedures for sourcing, evaluating, and integrating salvaged concrete slabs from buildings scheduled for demolition into a prototype of a novel structural system for application as load bearing systems in multi-story buildings.

Initially, a detailed examination of the concrete structures slated for demolition is conducted. If the mechanical properties, geometric characteristics, and overall condition of the building structure meet the required specifications, the process of harvesting the components is initiated. When working with reused materials, the design and structure of the developed system should be guided by the unique characteristics of the harvested components. To conceptualize an efficient and sensible design, the new structure must be closely aligned with the specific properties of the available concrete slabs, ensuring both architectural quality and structural integrity [20]. Table 1 illustrates the stages and processes involved in the construction of the reuse prototype.

To facilitate future reusability and enable multiple life cycle application for the components, the reused slabs are incorporated into a modular structural system. This system is designed according to the principles of DfD, employing newly developed mechanical connections that allow the slabs to be assembled and disassembled using mechanical force.

One of the biggest challenges when incorporating vast amounts of reused concrete slabs from different doner buildings is managing their unique geometries. While the length and width of a slab can often to some extent be adjusted during the deconstruction process, the thickness cannot be modified. To address this issue, the system's design must allow for maximum flexibility when accommodating elements of varying thicknesses. Another aspect addressed is the tectonic logic within the system. To figure out the most efficient approach, two construction designs are tested. The first one places the beam on top of the reused column, the second one inserts the beam into the reused column. By constructing both designs in real-world conditions,

Table 1 Stages within the reuse process for concrete elements

	Stage	Description
1.	Identification	Searching for a building that fits the requirements
2.	Assessment	Identifying the structural logic and ramifications
3.	Controlled deconstruction	Careful dismantling of the donor building
4.	Adaptation and repair	Modifications to fit new structural requirements
5.	Storage	Storage of the components until the start of the building site
6.	Reconfiguration	Assembly into the new system with DfD connections

the advantages and disadvantages of each variation can be clearly observed and recorded.

To evaluate the environmental performance of this prototype, the GWP is calculated and compared to traditional concrete systems as well as conventional DfD systems made with new materials.

2.2 Sourcing

The sourcing of reusable concrete elements presents several logistical challenges due to the scarcity of detailed information regarding the condition of buildings designated for demolition. In this paper, sourcing began by identifying potential buildings through information provided to us by local demolition companies. Suitable components had to meet the following criteria:

- The building must be scheduled for full deconstruction or demolition.
- Clear ownership of the concrete elements.
- Visual integrity of the slabs without major damage.
- Reinforcement bars must remain intact with no visible signs of corrosion.
- The concrete geometry must be compatible with the new design

After reviewing the available data and visiting different sites, two structures were chosen. The first batch of concrete elements was harvested from a reinforced concrete wall in St. Gallen. The second batch was also harvested in St. Gallen, but from a building currently undergoing renovation. Both dismantling projects follow the same steps, highlighted below.

1. Setting up the construction site and securing the perimeter
2. Cutting the concrete element
3. Drilling holes into the element to secure anchors for crane attachment
4. Connecting the steel hooks from the crane
5. Lifting the element and placing it onto a truck
6. Transporting it to a storage facility

Figures 1, 2 and 3 document the process of harvesting the first batch of concrete elements from the building site in St. Gallen. To be processed and adapted the parts are transported to their interim storage facility in Oberglatt, which constitutes a journey of 90 km.

2.3 Quality Inspection

Before the elements are sourced, an initial visual inspection is conducted. If no visible damage is found, the process continues. Once sourced, the elements undergo thorough quality inspections. Special attention is given to identifying potential cracks,

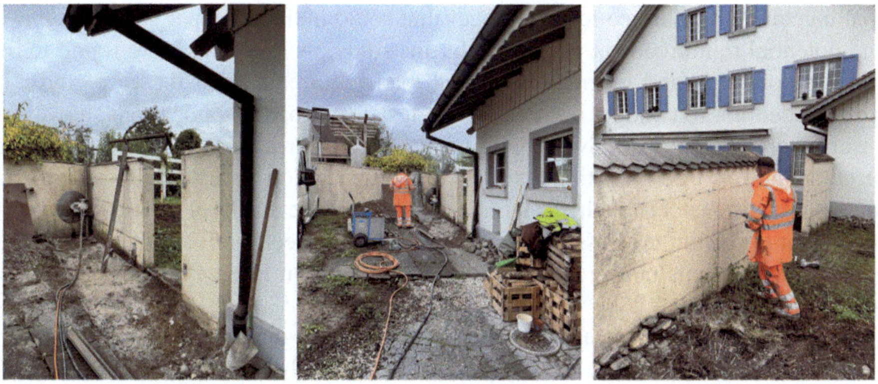

Fig. 1 The construction site is prepared, and the machinery is installed. After the element is cut, holes are drilled for the steel anchors, which are then placed into the holes

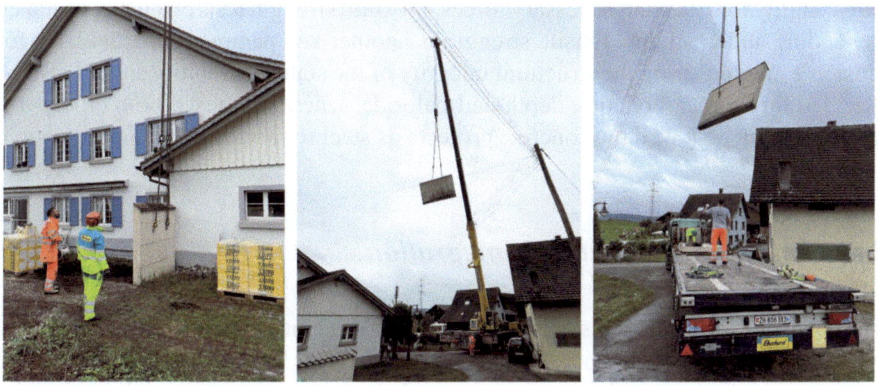

Fig. 2. A steel chain connects the concrete element to the crane, which then lifts the element onto a truck

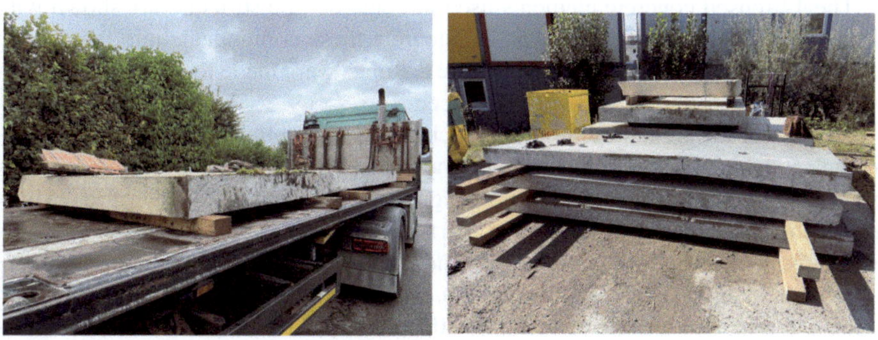

Fig. 3 The elements are secured on the truck and transported to the interim storage facility in Ober-Glatt, where they will undergo modification

spalling, and other surface-level damages that could indicate deeper structural flaws and potential reinforcement corrosion. In this study, the elements were found to be in good condition and given the minimal load requirements of the case study, it was decided not to conduct more thorough tests. In broader applications, however, ensuring compliance with Swiss building standards, particularly SIA 262 Concrete Structure and SIA 269 Existing Structures, is crucial. Non-destructive tests, such as ultrasonic pulse velocity, help detect internal defects like cracks or voids, while reinforcement corrosion tests assess the condition of the steel reinforcement. Concrete cover measurements ensure that the reinforcement is in line with local fire safety regulations.

In more demanding scenarios, cyclic loading and fatigue tests can be evaluated to show how the slabs perform under repeated loads, which can be important given the potential stress history of reused materials. Load testing provides a real-world assessment by replicating the conditions under which the slabs will function, ensuring they can bear the required loads. Additionally, compressive strength tests confirming the slabs' ability to withstand necessary forces. Flexural strength tests evaluate resistance to bending and cracking. Tensile strength is another key parameter, particularly for tension zones, ensuring the structural integrity of the slabs. Durability should also be assessed through carbonation depth and chloride penetration tests, both crucial for understanding how well the concrete protects its steel reinforcement from corrosion.

2.4 Design and Structural Integration

The design of the structural system leverages the structural integrity and mechanical properties of the reused concrete slabs. Since concrete undergoes carbonation and gains compressive strength over time, the goal was to develop a system that primarily utilizes the compressive strength of the elements. Figure 4 shows the reuse elements within their designated position in the overarching construction system.

The 14 reused concrete elements used to construct the structure are shown with their modification details in Fig. 5. To integrate these components into the design, adaptations are required. A total of 49 core drillings were made to create round holes where the steel elements will later be placed.

To explore different structural approaches and assess the architectural, environmental and logistical ramifications of the reused elements, two different beam-bearing methods are tested. These design alternatives aim to determine which approach better balances structural performance and architectural expression. Elements numbered 3 and 4, as well as 6 and 7 in Fig. 5, demonstrate these variations.

In the first design, used for the column constructed from parts 3 and 4, the beam is secured on top of the reused column with a mandrel. This method allows for clear load transfer directly onto the column while maintaining the integrity of the reused slab.

The second design, used for parts 7 and 8, takes a different approach by placing the beam within a cutout area in the reused column. This method involves creating

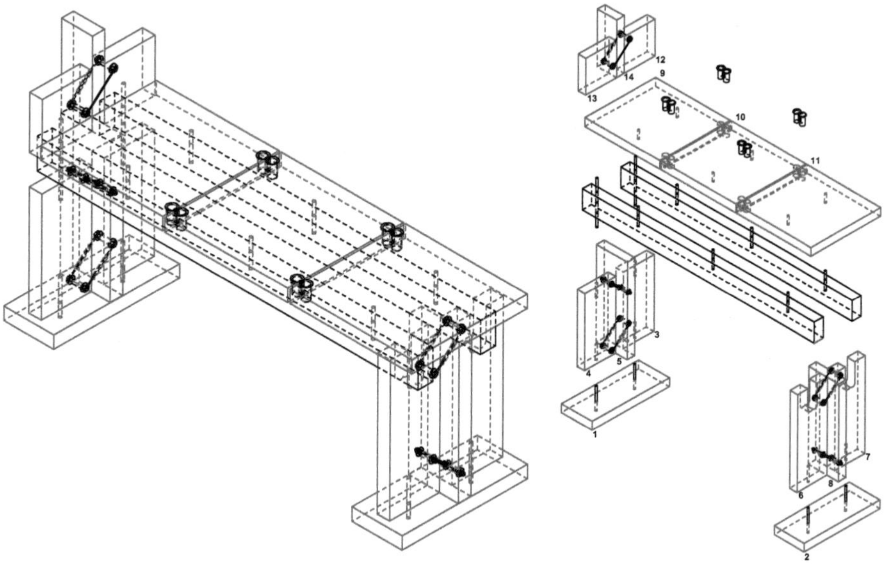

Fig. 4 Axonometric view of the structure left shows the assembled structure, right shows the exploded view. The reused elements are colored in green, while new elements are colored in black

two 65 cm deep cuts in both elements and using overlapping core drillings to separate the part. The cutout allows the beam to be embedded into the column, potentially providing enhanced lateral stability. In this case, the beam support will be leveled with a layer of mortar. In this design, no mandrel is used to connect the components. Instead, the elements rely on the clinched, mortared connection to form a weight-bearing joint.

These two approaches are employed to assess the advantages and disadvantages of each configuration. The first design allows for simpler construction and assembly, potentially reducing labor and time on-site, while the second design may offer improved integration and aesthetic coherence within the structure, aligning more closely with DfD principles. These tests allow for an evaluation of both approaches in terms of their structural capacity and architectural flexibility, guiding decisions for broader applications of reused materials in future projects.

Starting with the lowest positioned components in the scheme, each element is placed one on top of the other. The baseplate of the 1:1 mockup, which is also composed of two reused elements, is positioned first. Steel mandrels which are mortared in the baseplate secure the reuse columns which are placed on top.

The new load-bearing columns are created by placing three concrete slabs together, with the middle slab rotated 90 degrees relative to the others which can be seen in Fig. 6. Newly fabricated steel sleeves are inserted into the concrete slabs, enclosing the middle piece. Figure 7 shows the dimensions of the steel connections and threaded rods, illustrating how the nuts are used to lace the reused elements together, forming a rigid structural system. The manufactured steel elements can be

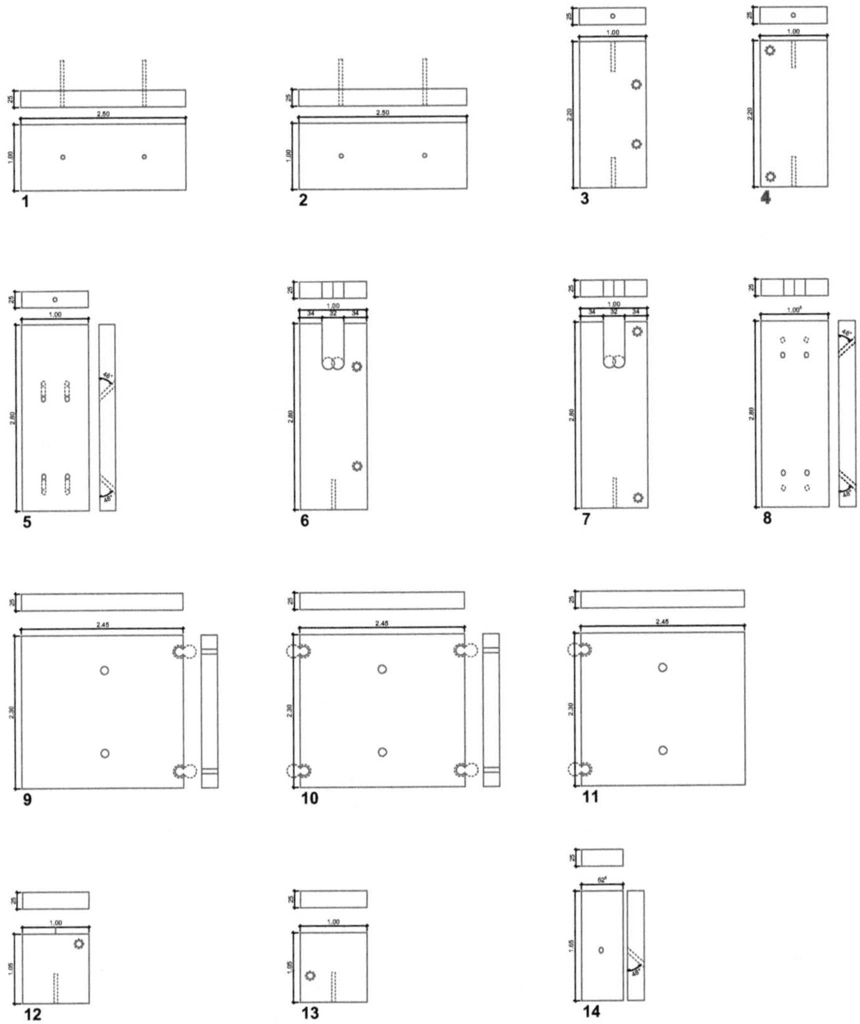

Fig. 5 Array of the reuse components being utilized with in the design of the structure. The reused elements are colored in green, while new elements are colored in black

seen in Fig. 8. A key design feature of the steel elements is their ability to integrate concrete slabs of varying thicknesses. This is made possible by the design of the connection, where two steel sleeves slide into each other and can be adjusted to accommodate different slab thicknesses. This flexibility significantly expands the pool of available parts that can be reused, making it easier for the construction industry to adopt and implement reused materials in various projects.

To bridge various distances, this structural system incorporates newly fabricated concrete beams. These beams will be placed either on top of or between the laced

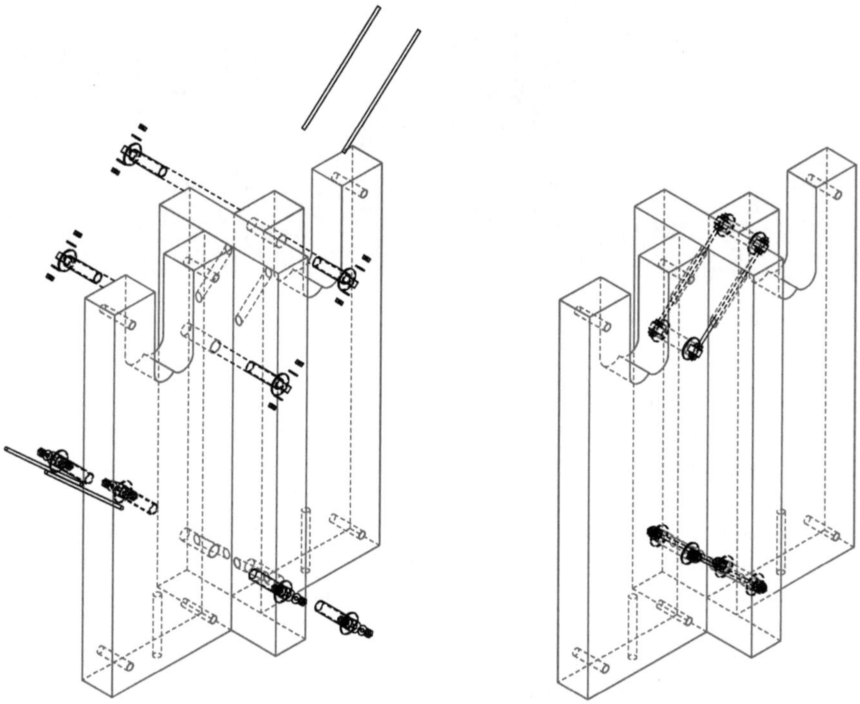

Fig. 6 Scheme of the lacing of the concrete column. The reused elements are colored in green, while new elements are colored in black

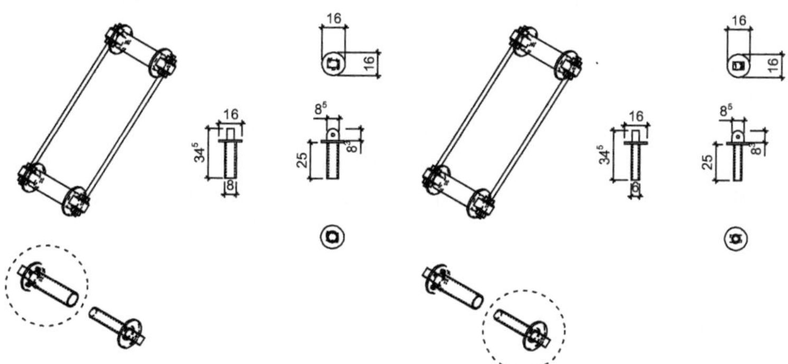

Fig. 7 Detail of the steel connections and how they come together

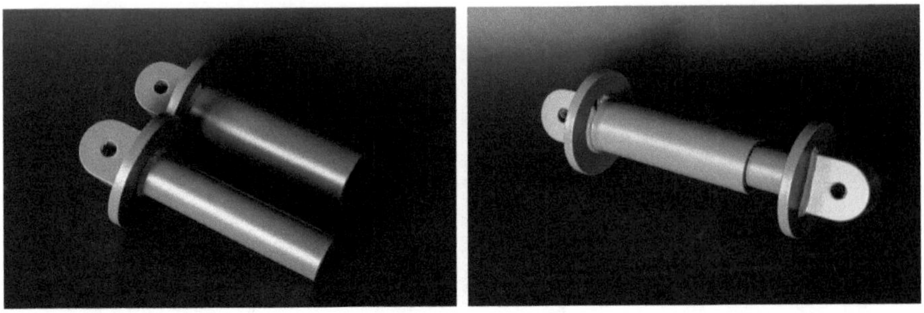

Fig. 8 Photodocumentary of the manufactured steel part

reused concrete columns. The performance of the structural system can be adjusted by altering the span width of the newly produced beams, allowing for flexibility in different applications. The beams support another set of reused concrete elements, which form the ceiling system. To create a stiffened structure, the elements are connected using custom-made steel components, as shown in Fig. 9. These connections utilize the core-drilled holes to anchor the sleeves in place. Figure 10 outlines the design and dimensions of the sleeve, while Fig. 11 shows the manufactured element. To accommodate different element thicknesses within the ceiling system, the steel sleeves are fitted with overlapping metal plates. This design ensures that the connection remains secure despite variations in slab thickness. The insertion depth of the sleeve does not have to be up to its entirety to achieve a weight-bearing connection, further enhancing the flexibility and ensuring compatibility with slabs of varying dimensions. The system design includes a gap between the elements to accommodate potential surface irregularities, which will be filled with mortar to achieve a level surface.

To demonstrate the continuity of the system, a shortened version of the reused concrete column will be placed upon the one-story high column.

The design adheres to principles of circularity, with specific attention given to creating a structural system that could be dismantled and reassembled without damaging the concrete slabs or other parts within the structural system.

2.5 Environmental Assessment

To assess the environmental performance of the reused concrete slabs a LCA in accordance with SN EN 15804 standards is performed [19]. For comparison, two design alternatives are defined. The same structure, made from conventional reinforced concrete will be evaluated both in a traditional construction method and in a DfD approach utilizing the manufactured components presented within this paper. The analysis covers stages A1–A3, as outlined in SN EN 15804, from the deconstruction of the original building to the reuse of the slabs in the new structure, phase

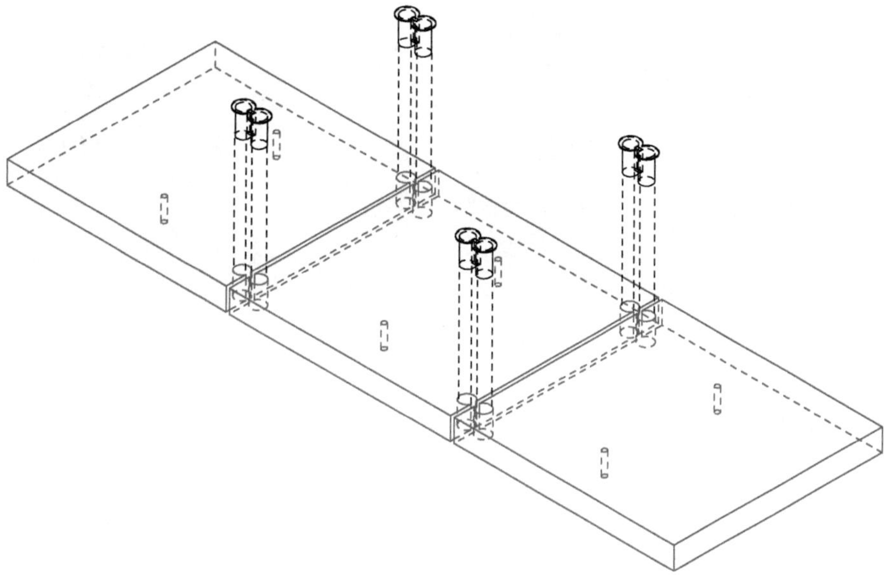

Fig. 9 Scheme of the sleeve connection the reuse concrete slaps. The reused elements are colored in green, while new elements are colored in black

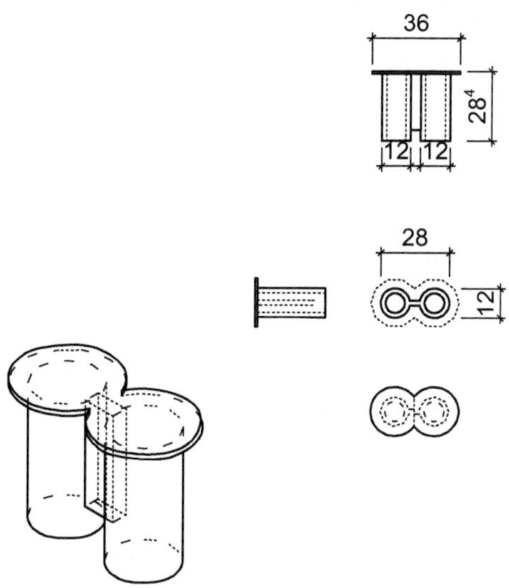

Fig. 10 Detail of the steel sleeve used to connect the reuse concrete slaps

R1–R3, accounting for deconstruction, transport, and modification. The LCA data is derived from the Swiss KBOB database, incorporating the GWP of each phase [20]. The results are presented in Table 2.

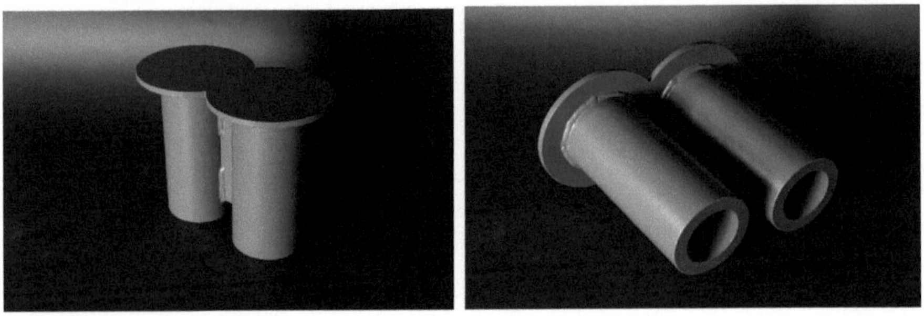

Fig. 11 Photodocumentary of the manufactured steel sleeve

3 Results

The GWP of the Reuse Concrete System amounts to a total of 2947.82 kg CO_2-eq, with the alternative, Traditional Concrete System amounting to 7889.45 kg CO_2-eq and the Traditional Concrete System DfD amounting to 9173.55 kg CO_2-eq. Compared to newly produced concrete, the reuse construction demonstrated a reduction in GWP by approximately 62.6, and by 67.9% when compared to the DfD design alternative. This significant decrease can be attributed to the avoidance of newly produced concrete, which is a major contributor to carbon emissions.

The environmental savings in the Reuse Concrete System were slightly offset by the logistical challenges of harvesting, transporting, and modifying the elements. These activities accounted for 7.9% of the total GWP, making them a minimal factor in the overall environmental impact.

4 Discussion and Future Work

One of the biggest challenges when working with reused materials is their sourcing. Within this paper, the parts were sourced through a company specialized in modifying existing concrete buildings and elements.

Reflection on the design variations tested, it is fair to say that each offer different advantages and disadvantages. Placing the beam on top of the reused column and securing it with a mandrel simplifies the assembly process but leaves a 60 cm gap that would need to be filled when the system is used in the future. In contrast, the design alternative where the beam is inserted into a cutout within the reused column requires more steps and precise execution. However, this method ensures that all gaps are closed during assembly, eliminating the need for additional modifications or fillings in the future. These designs also explore different tectonic approaches, assessing not only structural performance but also architectural potential.

A special feature of the developed system is its adaptability, which is particularly beneficial when working with reused materials, allowing for the incorporation of

Table 2 Results of the LCA with CO_2 emissions and material volumes

Pos.	Element	Quantity unit (kg)	Manufacturing and disposal CO_2 eq. (kg)	R1 selective deconstruction CO_2 eq. (kg)	R2 transport CO_2 eq. (kg)	R3 preparation CO_2 eq. (kg)	CO_2 GWP eq. (kg)
1	Reuse concrete slap	26,946.00		55.3	152	25.4	232.70
2	Steel connection wall	105.34	472.98				
3	Steel connection slap	96.74	434.36				
4	Steel mandrel	77.56	348.24				
5	Mortar	108.00	28.51				
6	Concrete beam	6105.05	1245.43				
7	Reinforcement	122.101	185.59				
							2947.82
	Traditional concrete system						
1	Concrete	32,601.05	6650.61				
2	Reinforcement	815.0263	1238.84				
							7889.45
	Traditional concrete system DfD						
1	Concrete	32,601.05	6650.61				
2	Reinforcement	815.0263	1238.84				
3	Steel connection wall	105.34	472.98				

(continued)

Table 2 (continued)

Pos.	Element	Quantity unit (kg)	Manufacturing and disposal CO_2 eq. (kg)	R1 selective deconstruction CO_2 eq. (kg)	R2 transport CO_2 eq. (kg)	R3 preparation CO_2 eq. (kg)	CO_2 GWP eq. (kg)
4	Steel connection slap	96.74	434.36				
5	Steel mandrel	77.56	348.24				
6	Mortar	108.00	28.51				
							9173.55

slabs of varying thicknesses. Utilized through the newly manufactures steel connections, this flexibility allows different reused elements to be effectively integrated into new designs while maintaining both structural integrity and disassemble ability.

This research highlights the potential for unconventional reuse strategies to significantly reduce greenhouse gas emissions in the construction industry. Focusing on the reuse of structural concrete elements, this study opens new pathways for more sustainable building practices, showing that salvaged materials can not only meet structural requirements but also outperform conventional systems in terms of environmental impact. Additionally, it demonstrates the feasibility of incorporating slabs of varying thicknesses while maintaining flexibility and reusability through a fully disassemble system.

5 Conclusions

This paper demonstrates the significant environmental benefits of reusing concrete elements to create a new prototype of a structural system for the construction industry, realized through a 1:1 mockup. The reuse of concrete slabs in this study resulted in 4941.63 kg CO_2-eq emissions, leading to a savings of 6225.73 kg CO_2-eq when compared to conventional construction methods. The comparative LCA showed that the reused concrete system had the lowest GWP, outperforming conventional concrete systems by over 62.6% and conventional DfD systems by 67.9%.

One of the key challenges in material reuse is the systematic sourcing and procurement of concrete slabs that meet specific geometric and structural requirements. In this paper, the identification of suitable elements was conducted through pragmatic means, such as collaboration with local companies and on-site inspections. However, more efficient methods, such as the creation of an open-source database of available decommissioned buildings and reusable materials could significantly streamline this process.

Another challenge is the logistical difficulty of extracting and transporting reuse concrete, particularly when traditional deconstruction methods that destroy the components are quicker and more economically viable. Future research must focus on optimizing the process of identification, data gathering, and controlled deconstruction of buildings marked for demolition.

The next phase of this research involves further development the presented structural system to address building physics and fire safety regulations, thereby maximizing the potential of this approach. Additionally, the material quantities must be reassessed and optimized, particularly concerning the amount and dimensions of the steel connections and new beams. Careful refinement is necessary to enhance both efficiency and the economic ramifications connected with this approach.

Future work should aim to optimize and expand this approach, including testing additional typologies and refining the use of materials such as the dimensions of steel connections and beams. By further enhancing material efficiency and reducing costs, this approach can drive forward the adoption of circular construction principles across the industry, helping to transform how we design, construct, and deconstruct buildings.

References

1. Scrivener, K.L., John, V.M., Gartner, E.M.: Eco-efficient cements: Potential, economically viable solutions for a low- CO_2, cement-based materials industry. Cement and Concrete Research 114, 2–26 (2018).
2. Mehta, P.K., Monteiro, P.J.: Concrete: Microstructure, Properties, and Materials. McGraw-Hill Education, New York (2014).
3. Andrew, R.M.: Global CO_2 emissions from cement production. Earth System Science Data 10(1), 195–217 (2018).
4. Habert, G., et al.: Environmental impacts and decarbonization strategies in the cement and concrete industries. Nature Reviews Earth & Environment 1(11), 559–573 (2020).
5. Cement Sustainability Initiative (CSI): The Role of Cement in the 2050 Low Carbon Economy. World Business Council for Sustainable Development, Geneva (2018).
6. Graser, J., Staufer, A., Meier, C. (eds.): Architektur Klima Atlas: Klimabewusst entwerfen in Forschung, Lehre und Praxis. Park Books, Zürich (2024). ISBN 978-3-03860-303-0.
7. Waldman, B., et al.: Embodied carbon in construction materials: A framework for quantifying data quality in EPDs. Buildings and Cities 1(1), 625–636 (2020).
8. Chastas, P., et al.: Embodied energy in residential buildings – towards the nearly zero energy building: A literature review. Building and Environment 105, 267–282 (2018).
9. Gorgolewski, M.: Designing with Reused Building Components: Some Challenges. Building Research & Information 36(2), 175–188 (2008).
10. Densley Tingley, D., Davison, B.: Developing an LCA Methodology to Account for the Environmental Benefits of Design for Deconstruction. Building and Environment 57, 387–395 (2012).
11. Ellen MacArthur Foundation: Towards the Circular Economy: Economic and Business Rationale for an Accelerated Transition. Ellen MacArthur Foundation, Cowes (2013).
12. Pomponi, F., Moncaster, A.: Circular economy for the built environment: A research framework. Journal of Cleaner Production 143, 710–718 (2017).
13. Flipsen, B., Bakker, C.A., de Pauw, I.C.: Examining the Role and Future Potential of Design for Disassembly Methods to Support Circular Product Design. In: Proceedings of the Design Society, vol. 3, pp. 1735–1744. Cambridge University Press, Cambridge (2023). https://doi.org/10.1017/pds.2023.174
14. Grošelj, P., et al.: Architectural Perspectives on Wood Reuse within Circular Construction: A South–Central European Study. Buildings 14(3), 560 (2024). https://doi.org/10.3390/buildings14030560
15. Cervantes Puma, G.C., Salles, A., Turk, J., Ungureanu, V.: Utilisation of Reused Steel and Slag: Analysing the Circular Economy Benefits through Three Case Studies. Buildings 14(4), 979 (2024). https://doi.org/10.3390/buildings14040979
16. Kanyilmaz, A., Birhane, M., Fishwick, R., et al.: Reuse of Steel in the Construction Industry: Challenges and Opportunities. International Journal of Steel Structures 23, 1399–1416 (2023). https://doi.org/10.1007/s13296-023-00778-4
17. Kavoura, F., Veljkovic, M.: Design Strategies and Technical Practices for Reusable Steel–Concrete Composite Structural Systems. In: Bragança, L., Cvetkovska, M., Askar, R., Ungureanu, V. (eds.) Creating a Roadmap Towards Circularity in the Built Environment. Springer, Cham (2024). https://doi.org/10.1007/978-3-031-45980-1_4
18. Stricker, E., Brandi, G., Sonderegger, A., Angst, M., Buser, B., Massmünster, M.: Bauteile wiederverwenden: Ein Kompendium zum zirkulären Bauen. Park Books, Zürich (2021). ISBN 978-3-03860-259-0.

19. Swiss Association for Standardization (SNV). Sustainability of construction works—Environmental product declarations—Core rules for the product category of construction products (SN EN 15804:2012+A2:2019). 2012. Available online: https://www.snv.ch/en/standards (accessed on 20 September 2024).
20. KBOB. Ökobilanzdaten im Baubereich 2009/1:2022 v5.0 [Life Cycle Assessment Data in the Construction Sector 2009/1:2022]. Bern 2022. Available online: https://www.kbob.admin.ch/kbob/de/home/die-kbob/plattform-oekobilanzdaten-im-baubereich.html (accessed on 11 October).

Business Performance Effects of Circular Economy Implementation in Core and Non-core Activities

André Podleisek, Nicolas Hofer, Roman Hänggi, and Katharina Luban

Abstract This article analyzes how the organizational context of Circular Economy implementation is related to perceived business performance effects. Using a comparative analytical approach, the study distinguishes between companies that integrate Circular Economy practices into their core business activities and those that implement such practices in non-core areas. The results show that companies with core integration report positive effects more frequently across most business performance factors when compared to companies with non-core implementation. These differences are particularly evident for market-related, capability-oriented, and risk-related dimensions. A notable exception concerns cost-related performance dimensions. In comparison to other business performance factors, cost-related effects show a less pronounced dominance of positive effects and instead display a more balanced distribution of positive, neutral, and negative effects.

Keywords Circular business performance · Circular economy practice · Lean digital green · Survey

---

A. Podleisek (✉) · N. Hofer · R. Hänggi · K. Luban
IPEK Institute for Product Development, Industrial Engineering and Product Lifecycle Management, OST Eastern Switzerland University of Applied Sciences, Rapperswil, Switzerland
e-mail: andre.podleisek@ost.ch
URL: https://www.ost.ch/ipek

N. Hofer
e-mail: nicolas.hofer@ost.ch

R. Hänggi
e-mail: roman.haenggi@bluewin.ch

K. Luban
e-mail: katharina.luban@ost.ch

A. Podleisek et al. (eds.), *Advances in Circular Economy: Bridging Research and Practice*, Springer Proceedings in Earth and Environmental Sciences,
https://doi.org/10.1007/978-3-032-21547-5_15

1 Circular Economy in Companies: Research Background and Focus

Research on Circular Economy (CE) has expanded considerably in recent years, reflecting a growing body of academic work addressing alternative production and consumption models [1]. The existing literature covers a wide range of perspectives, from the definition and conceptualization [2] to the analysis of drivers and barriers to CE adoption at the firm level, as well as policy frameworks and institutional mechanisms intended to promote CE principles at national and international levels [3–8]. Several studies examine the extent to which CE principles are adopted in practice. In 2018, the OECD reported that adoption by companies remained limited and often partial, with only a small share of firms having implemented CE-related principles or business models [9]. At the same time, other studies indicate that a substantial number of companies were not familiar with the CE concept [10]. More recent research suggests that both, familiarity and adoption, have increased with time [11].

The decision to integrate CE practices into a company's business model is strongly influenced by economic considerations and can affect both strategic direction and operational processes. At the same time, external factors, such as institutional requirements or customer pressure, can also influence companies to adopt circular practices. However, when adoption is primarily reactive, initiatives are often implemented in a project-based manner and remain confined to non-core business activities. By contrast, when companies perceive clear value creation through CE, they are more likely to deliberately integrate it into their core business activities as part of a wider strategic commitment aligned with economic objectives and long-term value creation [12, 13]

Against this background, the present study examines whether the organizational integration of CE makes a difference for business performance factors. Specifically, it compares business performance effects between companies that integrated CE into their core business activities and those that implemented them in non-core activities. By focusing on this distinction, the study aims to clarify how different integration contexts can shape the direction of business performance outcomes and to support companies in assessing the strategic implications of CE implementation.

1.1 Study Design and Sample Characteristics

The empirical basis of this study is a Europe-wide survey conducted in December 2022, targeting companies from the industrial sector, with a particular focus on manufacturing companies. The survey was sent mainly to decision-makers and individuals in leadership positions. While the questionnaire includes a broad range of topics related to CE, the analysis presented in this paper draws on a subset of questions that focus specifically on items capturing perceived effects on business performance factors (BPF). Details on the entire set of questions, study design, methodology used for the questionnaire, and technical aspects of the survey are described in [11].

In total, 512 valid responses were collected. For the purpose of the analysis presented in this paper, attention is restricted to companies that reported an active implementation of CE. This includes respondents who indicated CE implementation either within their core business activities or within non-core activities. Companies that reported no implementation or expressed uncertainty or unfamiliarity with the concept are not considered further. This analytical focus results in a subsample of 351 companies with headquarters located in 28 different European countries. For the breakdown of the subsample, refer to Tables 1 and 2.

Table 1 Characteristics of the subsample

Attribute	Variable	n
Position	CEO/business owner	142
	Other C-suite officers	28
	Environmental/sustainability/governance manager	54
	Business development manager	49
	Other managerial positions	36
	Technical staff	23
	Other	18
	N/A	1
Economic activity	Manufacturer of durable products	142
	Manufacturer of non-durable products	71
	Manufacturer of parts or components	26
	Manufacturer of convenience products	4
	Repair and installation of machinery and equipment	7
	Retail/wholesale	22
	Service	49
	Other	30
Employee count	<10	79
	10–49	68
	50–249	57
	250–999	50
	1000–10,000	52
	>10,000	45
CE in	Core activity	216
	Non-core activity	135
Total (N)		351

Table 2 Company headquarter location

Country	n	Country	n
Germany	46	Spain	15
Switzerland	46	France	12
Netherlands	44	Austria	11
Italy	43	Belgium	10
Sweden	32	Denmark	8
United Kingdom	26	Other Europe	36
Finland	17	Others	5
		Total (N)	351

In the subset of questions related to BPF, respondents were asked to assess the direction of perceived effects rather than their magnitude. The effects were classified as positive, neutral, negative, and uncertain. This approach allows the analysis to focus on whether CE implementation in core versus non-core business activities is associated with favorable, unfavorable, or neutral business performance outcomes, rather than on the perceived strength of these effects.

2 Empirical Findings on Business Performance

The respondents that reported CE implementation were asked to evaluate its perceived impact on eight BPF. These factors capture key dimensions of economic and operational performance, including cost-related, market-related, technological, and resilience-oriented outcomes. The following sections present and discuss the empirical findings, with a particular emphasis on differences between CE implemented in core business activities versus non-core activities.

2.1 Material and Manufacturing Costs

Material and manufacturing costs represent a critical component of overall production expenses. These cost components may be influenced by CE-related changes such as shifts from primary to secondary materials, product design choices, and adaptations to manufacturing processes required to enable reuse, repair, or remanufacturing. Such changes affect material and manufacturing costs through altered input prices, additional processing requirements, or efforts related to redesign, testing, certification, and process adaptation. The perceived effects of CE integration on material costs are illustrated in Fig. 1.

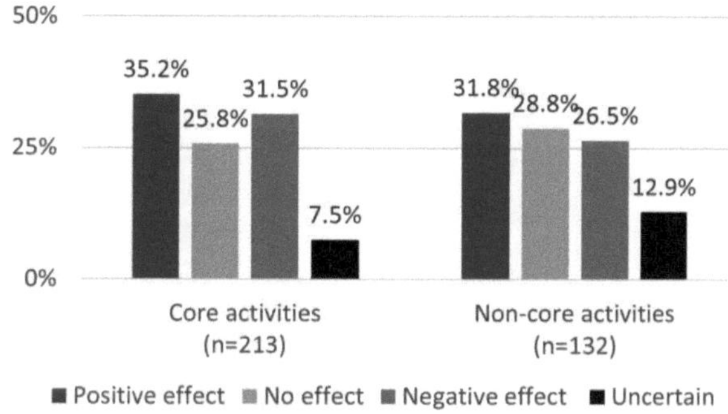

Fig. 1 Effects of CE on material costs

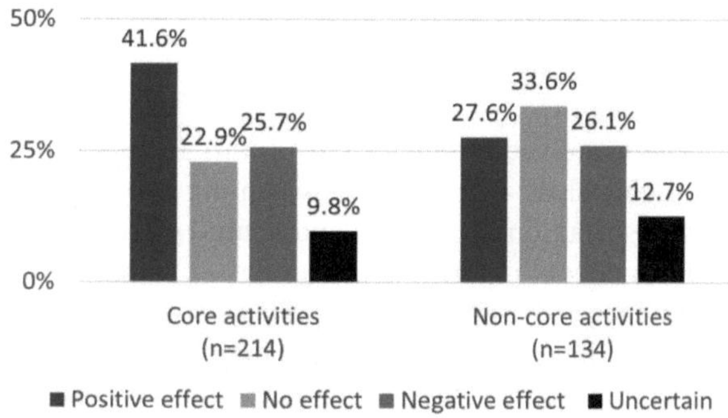

Fig. 2 Effects of CE on manufacturing costs

In the core business group, positive effects are reported by 35.2% of respondents, compared to 31.8% in the non-core group (difference: +3.4 percentage points). At the same time, negative effects are also reported on a slighty higher rate in the core group (31.5%) than in the non-core group (26.5%; difference: +5.0 percentage points).

A clearer distinction emerges with regard to manufacturing costs (see Fig. 2). Companies with CE integrated into core business activities reported positive impacts on a much higher rate than those with non-core implementation. In the core group, 41.6% of respondents report positive effects, compared to 27.6% in the non-core group (difference: +14.0 percentage points). At the same time, negative effects are reported by approximately one quarter of respondents in both groups. Neutral effects are reported more frequently in the non-core group, where 33.6% of the respondents also represent the largest share.

2.2 *Resilience Against Resource Related Risks*

Resilience against resource-related risks refers to a company's ability to cope with disruptions in the availability, price, or reliability of materials and resources. Such disruptions may arise from regulatory changes, geopolitical developments, increasing

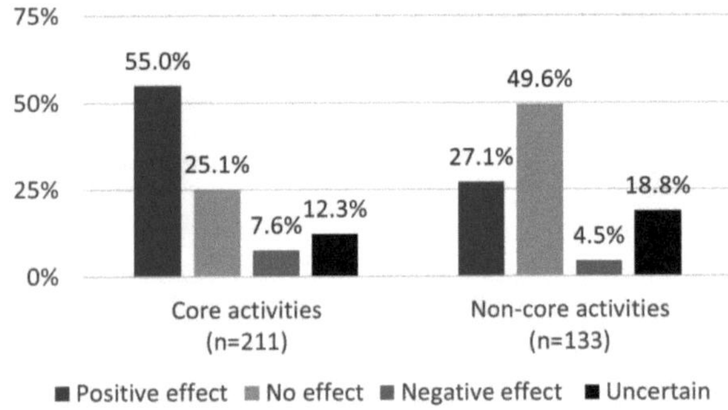

Fig. 3 Effects of CE on a company's resilience against resource-related risks

resource scarcity, or supply chain interruptions, as experienced during the COVID-19 pandemic [14].

The perceived effects of CE on this BPF are shown in Fig. 3. The comparison highlights clear differences between the two groups, primarily reflected in the relative distribution of positive and neutral effects. In the core business group, 55.0% of respondents report positive effects, compared to 27.1% in the non-core group (difference: +27.9 percentage points). By contrast, no percieved effect account for 25.1% in the core group, whereas they represent almost half of the responses in the non-core group (49.6%; difference: +24.5 percentage points). Negative effects on the company's resilience are reported by only a small share of respondents in both groups.

2.3 *Sales*

Sales performance reflects changes in revenue generation which can be associated with market acceptance, customer demand, and the commercial positioning of products and services. CE implementation can affect sales performance through product characteristics, service offerings, or communication related to circular attributes.

The perceived effects of CE on sales performance differ clearly between core and non-core implementation contexts, as shown in Fig. 4. Among companies implementing CE in core business activities, 73.4% report a positive effect on sales performance, while 15.0% indicate no effect and 2.3% report negative effects. In contrast, companies implementing CE in non-core activities report positive effects less frequently (39.6%), while neutral effects account for 47.8% of responses and negative effects for 1.5%. The distribution of effects indicate that positive effects on sales performance are much more likely when CE is implemented in core business activities than when it is limited to non-core activities.

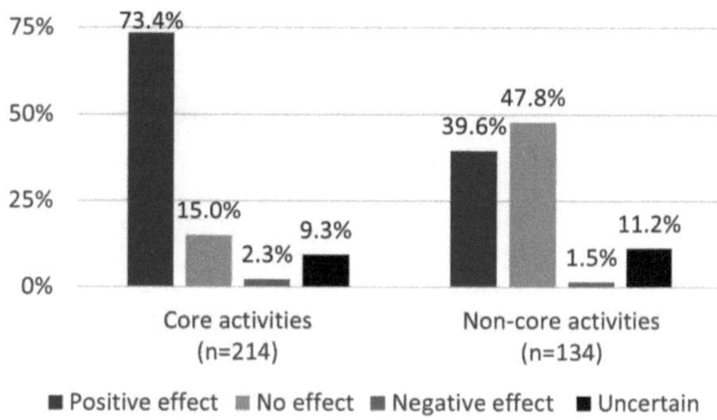

Fig. 4 Effects of CE on sales

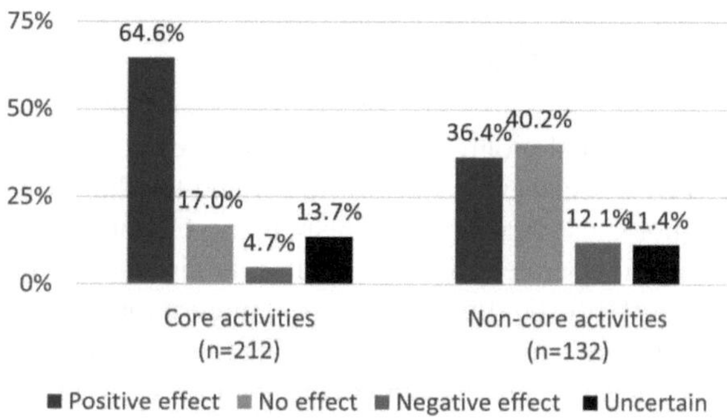

Fig. 5 Effects of CE on profits

2.4 *Profit*

Profitability reflects a company's financial success and captures whether revenues exceed costs over a given period. Figure 5 illustrates the perceived effects of CE on profitability. In the core business group, 64.6% of respondents report a positive effect on profitability, compared to 36.4% in the non-core group (difference: +28.2 percentage points). Neutral effects are reported by 17.0% of respondents in the core group, while this share increases to 40.2% in the non-core group (difference: +23.2 percentage points). Negative effects are reported by 4.7% of respondents in the core group and by 12.1% in the non-core group (difference: +7.4 percentage points).

In line with the pattern observed for sales performance, a distinct shift can be observed in the distribution of profitability-related effects attributed to CE between the two groups. Positive effects clearly dominate when CE is integrated into core business activities, whereas the non-core group is characterized by a substantially higher share of neutral effects. In addition negative effects occur more than twice as frequently in the non-core group compared to the core group.

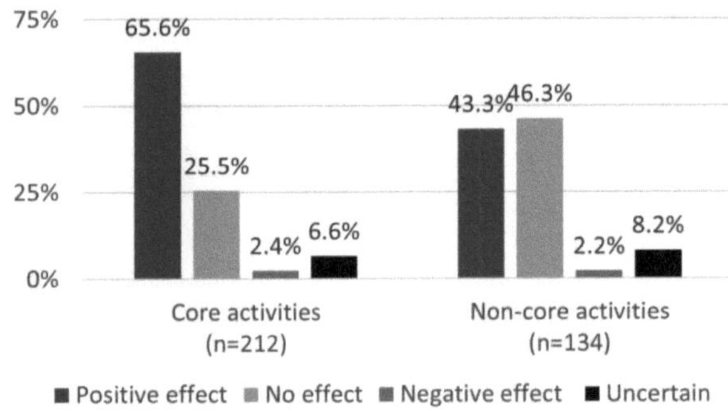

Fig. 6 Effects of CE on technological progress

2.5 *Technological Progress*

Technological progress focuses on changes in technological capabilities. It is treated as a BPF because it reflects whether and to what extent companies develop or adapt technological capabilities in response to CE integration. This can directly affect their ability to operate, compete, and scale circular activities. Examples from practice, such as DESSO, illustrate how CE integration can be associated with technological adaptations in areas such as material recovery and product redesign [15]. On the other hand, negative effects can occur when necessary CE-related technological changes are perceived as less suitable, less efficient, or more constraining than other existing technologies.

The results on this BPF are illustrated in Fig. 6. In the core business group, 65.6% of respondents indicate positive technological effects, whereas this share decreases to 43.3% when CE is implemented in non-core activities (difference: +21.7 percentage points). By contrast, the perception of no effect is stronger in the non-core group, where neutral assessments account for a substantially larger share of responses compared to the core group. Negative effects remain marginal in both implementation contexts.

2.6 *Waste Reduction*

As waste prevention and reduction are core elements of CE principles, and because they reflect operational efficiency outcomes that affect material use, process performance, and resource handling at the company level, waste reduction is treated as a business performance factor. As such, it captures a company's ability to manage resources efficiently and translate circular practices into measurable operational outcomes.

Figure 7 shows the distribution of perceived waste-related effects associated with CE implementation in core and non-core activities. In both groups, positive effects account for the majority of the responses, indicating that waste reduction is widely

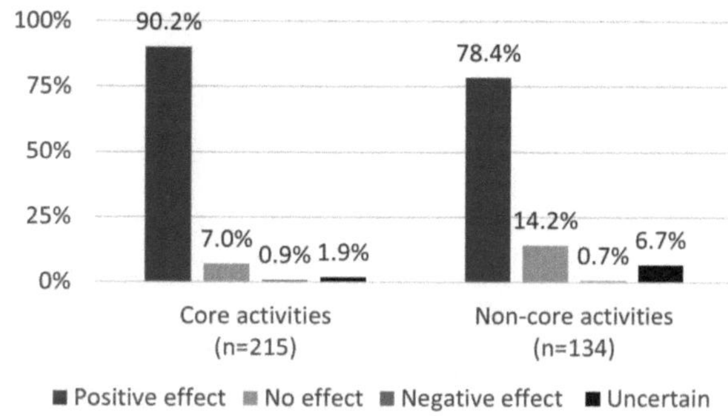

Fig. 7 Effects of CE on waste reduction

perceived as an outcome related to CE practices. The share of positive responses reaches 90.2% in the core business group and remains high at 78.4% in the non-core group. Differences between the two contexts are primarily reflected in the proportion of neutral effects (7.0% in the core group versus 14.2% in the non-core group] difference: +7.2 percentage points). Negative effects remain marginal in both groups (below 1%). A share of 6,7% of respondets in the non-core group reports uncertain effects, this suggests that waste-related effects are not systematically captured or assessed, for instance because corresponding data are not collected or not available in this implementation context.

Compared to other business performance factors, waste reduction shows high levels of positive effects across both implementation contexts.

2.7 Company Image

The final BPF considered in this analysis, company image, reflects how a firm is perceived by external and internal stakeholders, such as customers, employees, and investors. As a BPF, company image captures perception-based outcomes that can influence trust, legitimacy, and stakeholder support. In the context of CE implementation, this BPF focuses on whether CE-related activities are perceived as shaping a company's reputation and credibility.

Across both implementation contexts, CE is widely perceived as having a positive influence on company image (see Fig. 8). In the core business group, 93.0% of respondents report a positive effect on company image. A similarly positive assessment is also evident in the non-core group, where 76.9% of respondents indicate a positive effect. Notably, no negative effects on company image are reported in either group. Differences between the two contexts are therefore primarily reflected in the share of neutral assessments, which occur more frequently when CE is implemented in non-core activities.

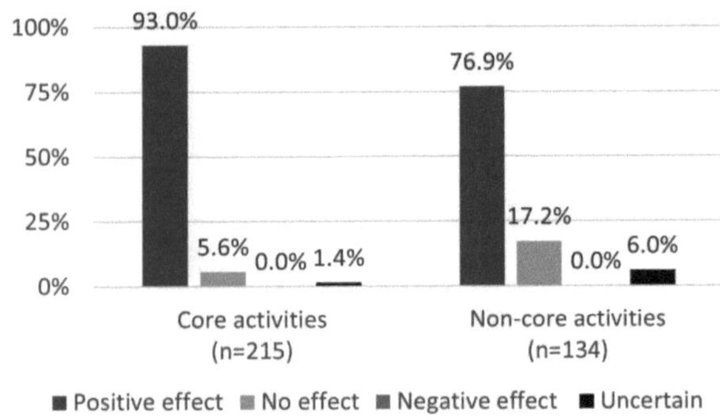

Fig. 8 Effects of CE on company image

3 Synthesis of Findings

The comparison between CE implementation in core and non-core business activities reveals distinct effect profiles that cut across cost-related, market-related, operational, and perception-based factors.

A first pattern emerges for market-oriented BPF, including sales performance and profitability. For these, CE implementation in core business activities is associated with a clear dominance of positive effects, whereas non-core implementation is characterized by a pronounced shift towards neutral assessments. Negative effects remain but occur more frequently in the non-core context. This pattern suggests that the organizational placement of CE is closely linked to whether market- and revenue-related effects are perceived as material.

A second pattern can be observed for resilience against resource-related risks and technological progress. Very, similarl to sales and profitability, these factors exhibit substantially higher shares of positive effects when CE is integrated into core business activities. The dominant contrast between the two implementation contexts also lies in the prevalence of neutral assessments in the non-core group. This indicates that CE-related contributions to resilience and technological capabilities are less clearly perceived when CE remains confined to non-core activities.

Cost-related factors show a more differentiated pattern. Material and manufacturing costs show mixed distributions of positive and negative effects in both groups, with comparatively smaller differences between core and non-core implementation. This ambivalence suggests that cost-related effects are not uniformly associated with CE integration and may depend on context-specific conditions, such as material choices, process configurations, or transition phases.

In contrast, waste reduction and company image form a distinct group of BPF. For both of them, positive effects dominate across core and non-core implementation contexts, and negative effects are either marginal or entirely absent. Differences between the two groups are primarily reflected in the share of neutral or uncertain assessments, particularly in the non-core group. This pattern indicates that some CE-related outcomes are widely perceived as positive irrespective of organizational embedding, while differences emerge mainly in the clarity and visibility of these effects.

Across all business performance factors, a recurring feature of the empirical results is the presence of uncertainty about the effects. The existence of these responses shows that, for a subset of companies, CE-related effects are not or cannot be clearly evaluated. The assessment of BPFs presupposes the availability of relevant data or reference values. Where such data are not systematically collected or monitored, companies are unable to compare performance before and after CE implementation. Beyond internal assessment, the absence of reliable data also limits the ability to communicate CE-related performance externally, making it difficult to signal whether and to what extent CE integration has delivered measurable benefits. These observations point to measurement, data availability, and transparency as important cross-cutting issues that shape how CE-related effects are evaluated and made visible.

3.1 Recommendations for Companies

The results support the view that CE is most effective when it is embedded within established operational excellence frameworks rather than pursued as a standalone sustainability initiative. Contemporary industrial operational performance is largely based on Lean principles and is increasingly enhanced through digitalization [16, 17]. Within this context, CE adds a complementary "green" dimension by extending efficiency-oriented operational models towards lifecycle-oriented and resource-focused value creation [18–20].

Implementing CE in this way requires an understanding of the entire product lifecycle and the ability to systematically use data to question and adapt existing operational routines. Combining Lean (stable processes), digitalization (data, technologies), and CE (sustainability) enhances operational performance. Based on these considerations, we propose a Lean–Digital–Green approach that integrates these three perspectives into a comprehensive operational excellence model (see Fig. 9).

The Lean–Digital–Green framework places the integration of CE practices within a broader operational excellence perspective. By combining Lean principles, digital

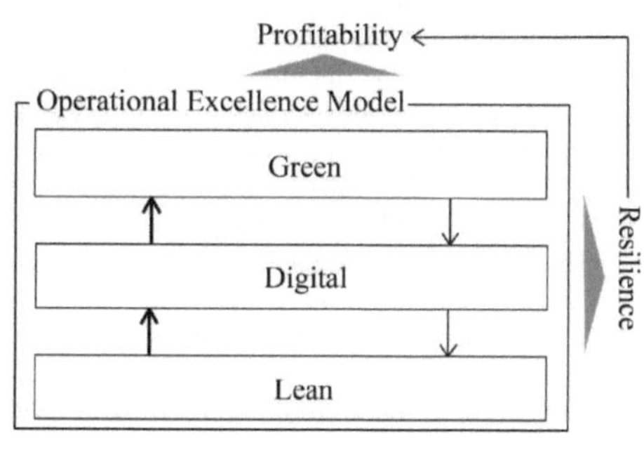

Fig. 9 Lean-digital-green framework for operational excellence

technologies, and sustainability-oriented thinking, the framework links process stability, data-driven transparency, and lifecycle-oriented resource management into a coherent approach. Lean provides the foundation for efficient and reliable processes, digitalization enables the systematic generation and use of operational data, and Circular Economy extends this foundation by introducing a lifecycle-oriented perspective on value creation. In this way, Lean–Digital–Green shifts attention from individual performance effects to the organizational conditions that enable CE implementation. The framework highlights how processes, data infrastructures, and responsibilities can be deliberately aligned to support CE initiatives and to make their effects observable, comparable over time, and communicable to internal as well as external stakeholders.

3.2 Further Research

While this study provides insights into the relationship between CE implementation and business performance, it is based on self-reported perceptions and cross-sectional data. Future research could explore causal relationships in greater depth and analyze interactions between different performance factors. Longitudinal studies and sector-specific analyses may further improve understanding of how CE contributes to sustainable competitive advantage.

References

1. Web of science, Search throughout All database using as Keyword Circular Economy from 1989 to 2022. (2022). https://www.webofscience.com/wos/woscc/analyze-results/8565a6aa-8d21-441b-81e1-e72cccb90ddf-68e87c00 (visited on 10/28/2022).
2. J. Kirchherr, D. Reike, and M. Hekkert, Conceptualizing the circular economy: An analysis of 114 definitions, Resources, Conservation and Recycling, vol. 127, pp. 221–232. (2017). https://doi.org/10.1016/j.resconrec.2017.09.005. (visited on 04/06/2022).
3. A. Podleisek, R. Hänggi, and K. Luban, Circular Economy: The Next Level of Company Success. Springer (2025).
4. J. Hart, K. Adams, J. Giesekam, D. D. Tingley, and F. Pomponi, Barriers and drivers in a circular economy: The case of the built environment, Procedia CIRP, vol. 80, pp. 619–624 (2019). https://doi.org/10.1016/j.procir.2018.12.015. (visited on 04/06/2022).
5. D. Reike, W. J. Vermeulen, and S. Witjes, The circular economy: New or refurbished as CE 3.0? Exploring controversies in the conceptualization of the circular economy through a focus on history and resource value retention options, Resources, Conservation and Recycling, vol. 135, pp. 246–264 (2018). https://doi.org/10.1016/j.resconrec.2017.08.027.
6. N. van Buren, M. Demmers, R. van der Heijden, and F. Witlox, Towards a circular economy: The role of dutch logistics industries and governments, Sustainability, vol. 8, no. 7, p. 647 (2016). https://doi.org/10.3390/su8070647. (visited on 05/10/2022).
7. T. Andersen, A comparative study of national variations of the European WEEE directive: Manufacturers view, Environmental Science and Pollution Research, vol. 29, no. 14, pp. 19920–19939 (2022). https://doi.org/10.1007/s11356-021-13206-z.

8. D. Vermunt, S. Negro, P. Verweij, D. Kuppens, and M. Hekkert, Exploring barriers to implementing different circular business models, Journal of Cleaner Production, vol. 222, pp. 891–902 (2019). https://doi.org/10.1016/j.jclepro.2019.03.052.
9. OECD, Business models for the circular economy: Opportunities and challenges from a policy perspective, OECD Publishing, Paris (2018).
10. D. Masi, V. Kumar, J. A. Garza-Reyes, and J. Godsell, Towards a more circular economy: Exploring the awareness, practices, and barriers from a focal firm perspective, Production Planning and Control, vol. 29, no. 6, pp. 539–550 (2018). https://doi.org/10.1080/09537287.2018.1449246.
11. N. Hofer, A. Podleisek, R. Hänggi, and K. Luban, Implementation of circular economy in European industrial companies a survey, in Circular Economy: The Next Level of Company Success, Cham: Springer, 2025, pp. 67–128.
12. K. Govindan and M. Hasanagic, A systematic review on drivers, barriers, and practices towards circular economy: A supply chain perspective, International Journal of Production Research, vol. 56, no. 1, pp. 278–311 (2018). https://doi.org/10.1080/00207543.2017.1402141.
13. M. A. Camilleri, European environment policy for the circular economy: Implications for business and industry stakeholders, Sustainable Development, vol. 28, no. 6, pp. 1804–1812 (2020). https://doi.org/10.1002/sd.2113.
14. K. Luban and R. Hänggi, Communicable Eds., Erfolgreiche Unternehmensführung durch Resilienzmanagement, Berlin, Heidelberg: Springer (2022)
15. Desso, As a cradle to cradle® company, we go beyond what most people understand by sustainability (2015). https://vloeren.projecten.tarkett.nl/nl_NL/node/onze-duurzaamheidsstrategie-omgezet-in-acties-1831 (visited on 04/14/2022).
16. R. Hänggi, A. Fimpel, and R. Siegenthaler, LEAN Production—Einfach und Umfassend: Ein Praxisorientierter Leitfaden zu Schlanken Prozessen mit Bildern Erklärt. Berlin, Heidelberg: Springer (2021)
17. L. Budde, R. Hänggi, T. Friedli, and A. Rüedy, Smart Factory Navigator: Identifying and Implementing the Most Beneficial Use Cases for Your Company—44 Use Cases That Will Drive Your Operational Performance and Digital Service Business. Cham: Springer (2023)
18. G. A. Queiroz, P. N. A. Junior, and I. C. Melo, Digitalization as an enabler to SMEs implementing lean-green? A systematic review through the topic modelling approach, Sustainability, vol. 14, no. 21, p. 14089 (2022). https://doi.org/10.3390/su142114089.
19. M. Roesch, J. Köberlein, M. Goldmanns, and A. Hohmann, Green-lean-digital als leitbild für die nachhaltige fabrik der zukunft, Zeitschrift für Wirtschaftlichen Fabrikbetrieb, vol. 116, no. 1, pp. 29–33 (2021). https://doi.org/10.1515/zwf-2021-0006.
20. A. Goelz, Green lean digital factory: Lean management trifft digitalisierung, 2020. https://industrie.de/industrie-4-0/green-lean-digital-factory-lean-management-trifft-digitalisierung/ (visited on 06/09/2023).

MIX
Papier aus verantwortungsvollen Quellen
Paper from responsible sources
FSC® C105338

If you have any concerns about our products,
you can contact us on
ProductSafety@springernature.com

In case Publisher is established outside the EU,
the EU authorized representative is:
Springer Nature Customer Service Center GmbH
Europaplatz 3, 69115 Heidelberg, Germany

Printed by Libri Plureos GmbH
in Hamburg, Germany